Climate Change Adaptation and Development

Climate change is real and man-made. We have put so many greenhouse gas pollutants into the atmosphere that we will see significant and long-term change that we need to adapt and adjust to. It is important for development practitioners to understand these impacts and the challenge of how and when to adapt to climate change.

There are plenty of grim presentations of what the extremes of the possible climate scenarios will throw at us over the next 100 years, but not all change will be disastrous; some change will be beneficial, but much of the change will happen at an unprecedented rate that will require the best possible analysis and understanding of how and when we should adapt to climate change.

This is important for development practitioners as we invest in ensuring that poverty is reduced and eliminated and the well-being of everyone is improved. Many countries and communities around the world are vulnerable to the impacts of climate change, but developing economies may on one hand be less resilient to the impact, but could on the other hand be in a better position to make their development climate smart by making the most efficient use of their economic resources.

The chapters in this book shine a light on the complexity and the multi-dimensional aspects of climate change adaptation. They gather some of the experiences of addressing climate change impacts in a development context. This book was previously published as a special issue of *Development in Practice*.

John Carstensen is the Head of Profession in Climate and Environment for the Department for International Development (DFID). He has over 30 years of experience in the field of sustainable development, environment and climate change from Government, United Nations and NGO perspectives. He led international negotiations to protect the Ozone Layer as the Chair of the Montreal Protocol working group from 1992 to 1995 and has delivered poverty-oriented environment programmes in Egypt, Thailand and Vietnam. He was the CEO of Society for the Environment and the Chief Operating Officer for INTRAC.

Climate Change Adaptation and Development

Edited by
John Carstensen

Routledge
Taylor & Francis Group

LONDON AND NEW YORK

First published 2017
by Routledge

2 Park Square, Milton Park, Abingdon, Oxfordshire OX14 4RN
711 Third Avenue, New York, NY 10017

Routledge is an imprint of the Taylor & Francis Group, an informa business

First issued in paperback 2018

British Library Cataloguing in Publication Data
A catalogue record for this book is available from the British Library

ISBN : 978-1-138-69607-5 (hbk)
ISBN : 978-0-367-02835-0 (pbk)

Typeset in TimesNewRomanPS
by diacriTech, Chennai

Publisher's Note
The publisher accepts responsibility for any inconsistencies that may have arisen
during the conversion of this book from journal articles to book chapters, namely
the possible inclusion of journal terminology.

Disclaimer
Every effort has been made to contact copyright holders for their permission to
reprint material in this book. The publishers would be grateful to hear from any
copyright holder who is not here acknowledged and will undertake to rectify any
errors or omissions in future editions of this book.

Contents

CONTENTS

Citation Information

The chapters in this book were originally published in the *Development in Practice*, volume 24, issue 4 (June 2014). When citing this material, please use the original page numbering for each article, as follows:

For any permission-related enquiries please visit:
http://www.tandfonline.com/page/help/permissions

Notes on Contributors

Simon Anderson is the former Head of the Climate Change group at IIED.

Diane Archer is a Researcher at the International Institute for Environment and Development (IIED) in London, UK.

Emily Benson leads communications for the Green Economy Coalition which is currently housed at IIED.

Robin Bloch is the Technical Director for urban planning at ICF International in London, UK.

John Carstensen is the Head of Profession for Climate and Environment at the UK Department for International Development (DFID).

Tiguist Fisseha is a Senior Disaster Risk Management Specialist at the World Bank, Washington, DC, USA.

Alex Forbes is a Programme Officer at PEI Africa.

Su-Lin Garbett-Shiels works at the Grantham Research Institute on Climate Change and the Environment, London School of Economics and Political Science, and the Department for International Development, London, UK.

Alex Harvey works at the Department for International Development, London, UK.

Ced Hesse is the Principal Researcher for Drylands, Climate Change Group and Team Leader on Research for Advocacy at IIED.

Muhammad Jahedul Huq is a PhD candidate at Durham University, UK. In the past he has worked as a Research Officer in Climate Change Adaptation and Disaster Risk Reduction with ActionAid Bangladesh. He has a Masters in Human Ecology from the Department of Human Ecology, Free University of Brussels, Belgium.

Saleemul Huq is a Senior Fellow with IIED and the Director of the International Centre for Climate and Development (ICCAD) at the Independent University in Dhaka. His areas of expertise lie in the interlinkages between climate change (both mitigation as well as adaptation) and sustainable development from the perspective of developing countries with special emphasis on the least developed countries.

Melanie S. Kappes is a Disaster Risk Assessment Specialist at the World Bank, Washington, DC, USA.

NOTES ON CONTRIBUTORS

Anup Karanth works for TARU Leading Edge in Gurgaon, India.

Fawad Khan is the founder of Institute of Social and Environmental Transition-Pakistan (ISET-PK.) He is also a Senior Associate at ISET–International and an International Fellow at International Institute of Environment and Development (IIED).

Mika Korkeakoski is the PEI Asia Pacific Programme Officer.

Andrea Kutter is a Senior Programme Coordinator (Forest Investment Programme and Pilot Programme for Climate Resilience) for the Climate Investment Funds of the World Bank, Washington, DC, USA.

Razi Latif is a climate and environment adviser in DFID who previously worked with PEI as the Asia Pacific region co-manager.

Dechen Lham was an intern with PEI in the Asia Pacific region in 2013.

Rawlings Miller is a Technical Specialist at ICF International, Boston, USA.

Marcus Moench has led ISET since its founding in 1997 and has guided the development of its programmes on urban resilience to climate change, water resource management, and disaster risk management ever since. He received his PhD from the University of California at Berkeley in 1989. He has published numerous articles and papers on natural resources management.

Jose Monroy is a Research Assistant at ICF International in London, UK.

Victor Orindi is a Climate Change Advisor at the Kenya National Drought Management Authority. He also coordinates the Adaptation Consortium.

Nikolaos Papachristodoulou is a consultant in the international development practice at ICF International in London, UK.

James Pattison works as a consultant for IIED, focusing on dryland livelihoods in East Africa.

Beatriz Pozueta is a Senior Disaster Risk Management Specialist at the World Bank, Washington, DC, USA.

Neha Rai is a researcher with Climate Change Group of IIED (International Institute of Environment and Development) in the UK. At IIED she is leading a study on understanding the political economy of Climate Investment Funds.

Nicola Ranger works at the Grantham Research Institute on Climate Change and the Environment, London School of Economics and Political Science, London, UK.

Bhaskar Reddy is a Special Secretary at the Planning and Coordination Department for the Government of Odisha in Orissa, India.

Niranjan Sahu is the GIS Manager at Orissa Watershed Development Mission in Orissa, India.

Janpeter Schilling is a Programme Officer in Climate Change and Security at International Alert in London and an Associated Researcher in the Research Group Climate Change and Security at Hamburg University, Germany. His field of specialisation is the linkages between climate change, vulnerability, adaptation, and conflict.

NOTES ON CONTRIBUTORS

Virinder Sharma is a Regional Climate Change Adviser at DFID Kenya and Somalia.

Dan Smith is the Secretary General of International Alert and Chairman of the UN Peacebuilding Fund's Advisory Group. Dan Smith also holds a professorship at the University of Manchester's Humanitarian and Conflict Response Institute.

Lorena Trejos is an Urban Development Specialist at the World Bank, Washington, DC, USA.

Janani Vivekananda has been the Environment, Climate Change and Security Manager at International Alert since 2009. Her specific interests include the implications of climate change policies on peace, the links between climate change and community resilience, and opportunities for positive responses to climate and environmental change and disasters.

Leon Dwight Westby is a Programme Analyst (Forest Investment Programme) for the Climate Investment Funds of the World Bank, Washington, DC, USA.

Tim Wheeler is the former Deputy Chief Scientific Adviser, Department for International Development, UK, and Professor of Crop Science, University of Reading.

INTRODUCTION

Climate change adaptation and development

John Carstensen, Editor

Climate change is real, it is happening, and it is man-made (IPCC WGI 2013). We also know that we have already put so many greenhouse gas (GHG) pollutants into the atmosphere that we will see significant and long-term change that we need to adapt and adjust to (IISD 2014). It is fundamentally important for all development practitioners to understand these impacts and to get to grips with the challenge of how and when to adapt to climate change.

As this special issue went to print, the international science community gathered to discuss their understanding of climate change impacts and to produce an update of the assessment that was made five years ago (IISD 2014).

There are plenty of grim presentations[1] of what the extremes of the possible climate scenarios will throw at us over the next 100 years, but not all change will be disastrous; some change will be beneficial, but much of the change will happen at an unprecedented rate that will require the best possible analysis and understanding of how and when we should adapt to climate change.

This is particularly important for development practitioners as we invest in ensuring that poverty is reduced and eliminated and the well-being of everyone is improved. Many countries and communities around the world are vulnerable to the impacts of climate change, but developing economies may on one hand be less resilient to the impact, but could on the other hand be in a better position to make their development climate smart by adapting at the right time and in the right places to make the most efficient use of their economic resources.

The articles in this special issue shine a light on the complexity and the multi-dimensional aspects of climate change adaptation. They gather some of the experiences of addressing climate change impacts in a development context. I would like to highlight a few facets of adaptation and development that are covered by these articles. These are aspects that I believe are important for actions and for the debate and discussion of priorities and approaches.

Dealing with uncertainty

When development practitioners are faced with decisions that involve adaptation to climate change we must balance a series of uncertainties. We do not know how effectively the world community will address the need to reduce GHG-emissions or how fast low carbon technologies[2] will be introduced – how well we, as a world, will be able to mitigate further climate change. We know that we are currently on a pathway slightly worse than the 'business as usual' scenarios(IPCC WGI 2013), but we also know that efforts to reduce emissions are accelerating in the industrialised countries and that growth in carbon emissions is slowing. This could be consistent with the reduction scenarios in the IPCC report on the physical climate science (RCP 2.6 and RCP 4.5), but only if additional and significant action is taken over the next few years (see Figure 1).

1

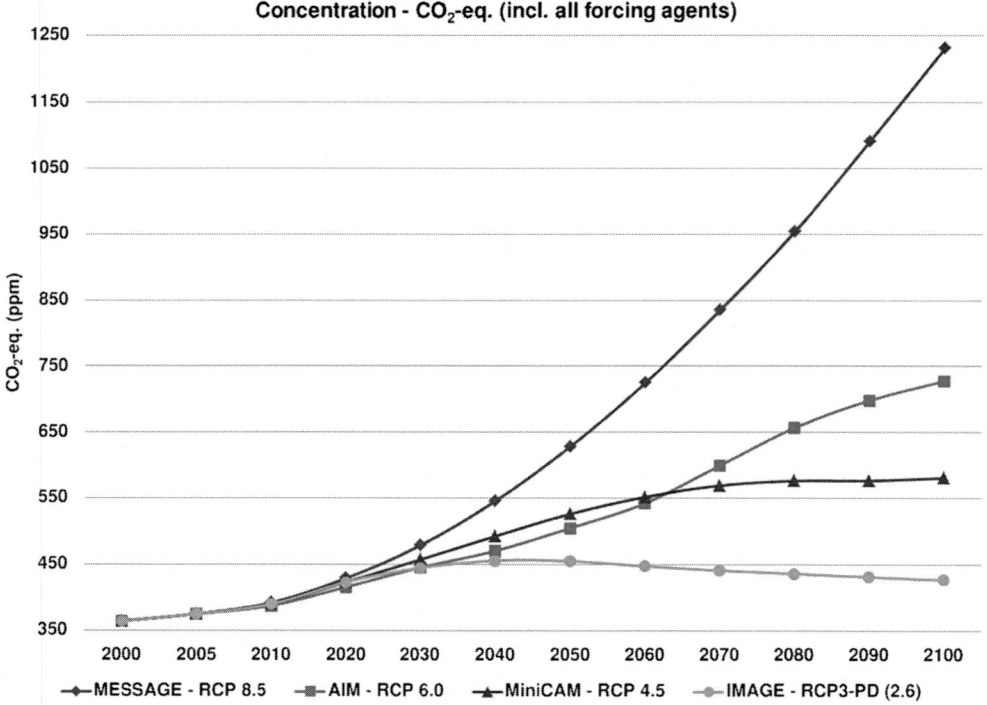

Figure 1.   Reduction scenarios

In addition, climate sensitivity is uncertain with temperature increases ranging from 1.5–4.5° C. This range was lowered slightly by the IPCC report in September 2013, compared to previous years' assessments. Finally, the global science community is just beginning to understand how nature and people react to such changes, and the most recent IPCC report on climate impacts was released as this special issue went to press. There is still a huge range in assessments of the economic impact of additional temperature increases of approximately 2°C, putting global aggregate economic losses somewhere between 0.2 and 2.0% of income. Losses will accelerate with additional warming. A recent World Bank (2013) report estimated global adaptation costs to be in the range of US$70–100 billion per year in developing countries from 2010 to 2050. This figure is often disputed, but probably still the most accurate estimate that exists. Both of these estimates are still incomplete and have recognised limitations. The headlines from the latest IPCC (2014) report (the Summary for Policy Makers of the Working Group II report on Impacts, Adaptation and Vulnerability) show that:

- This is the *most comprehensive and rigorous assessment* of the impacts of climate change ever produced, reflecting the consensus of hundreds of the world's leading scientists, based on more than 12,000 peer reviewed articles and agreed by more than 120 governments following an extensive review process.
- The impacts of climate change are expected to *slow down economic growth, erode food security, and make poverty alleviation more difficult.*
- *Unmitigated climate change poses great risks to global food and water security, human health, natural ecosystems, and economic development.* The poor and marginalised are most vulnerable. Urban areas, low-lying areas, and emerging hotspots of hunger are most at risk.

- Effective and inclusive climate change adaptation can help to build a richer, more resilient world in the near-term and beyond.
- Adaptation is essential in dealing with the risks of climate change, but there are limits to what adaptation can achieve, so urgent action is also needed to reduce emissions.
- Effectively dealing with the impacts of climate change is about good risk management.

Many of the articles in this special edition address parts of this complexity of uncertainty. Last year, DFID published a Topic Guide on decision-making under uncertainty (Ranger 2013) to help practitioners avoid the worst trap of all: to defer, avoid, or delay decisions because of the uncertainty of some aspects of change and adaptation.

To begin the issue, **Wheeler** sets out the imperative for investing in better climate adaptation and resilience as part of development assistance programmes. **Ranger et al.** look at the quality of those decisions by major funders and address how these organisations fare when it comes to assessments of the vulnerability of the people they provide assistance for.

Making good vulnerability assessments

With the uncertainties that exist, it can be a challenge to ensure good vulnerability assessments as a foundation for development interventions. Two articles look into the experience of vulnerability assessments in two different settings where only limited research has taken place: fragile and conflict settings and the emerging area of urban development. These pieces by **Vivekananda and Smith** and **Papchristodoulou et al.** both explore areas where practitioners are in need of more research, particularly as both are likely to be the target of more development interventions in the future.

Resilience

Three articles follow on from this to help us understand how building resilience to the impacts of climate change plays a central role in safeguarding development outcomes. **Moench** links the vulnerability assessments to a framework for analytical assessments and an iterative planning process developed for urban processes but with wider application for all climate resilience interventions. **Rai et al.** offer interesting insights from Bangladesh, one of the countries with perhaps the most developed experience in building resilience to weather variability and climate change.

The adaptation deficit

Khan's article on basic services for resilience opens an interesting angle that involves a rapidly-evolving concept: that of the adaptation deficit. This is the notion that interventions that need to be made irrespective of climate change impacts, either because they represent good practice in managing natural resources or a requirement for human well-being, will in themselves be effective measures to adapt to climate change and create resilience. The first paper by **Sharma et al** similarly explores how a focus on improving local governance systems, an important development goal outside climate change, creates the foundation for good climate adaptation. For many years, the joint UNEP/UNDP Poverty and Environment Initiative (PEI) has explored this area, and the paper by **Latif et al.** has collected useful learning from several areas covered by the programme.

Urban and rural

Several papers address the particular context of urban development, as mentioned above and the two contexts (urban and rural) in which interventions to build climate resilience take place are explored in three papers. **Kutter and Westby** look at the holistic landscape approach to ensuring well-integrated rural interventions. The second **Sharma et al.** paper also focuses on the experience in the rural context.

There are distinct differences in the approach needed in rural and urban climate change adaptation, and both are essential for a good understanding of effective climate resilience. **Karanth and Archer** offer a useful contribution to addressing these twin pillars of adaptation.

Notes

1. For example, see "Official prophecy of doom: Global warming will cause widespread conflict, displace millions of people and devastate the global economy", in the Independent Online, Tuesday 18 March 2014
2. Renewable energy, emission reduction technology, such as carbon capture and storage and energy efficiency measures.

References

International Institute for Sustainable Development (IISD). 2014. "Summary of the 10th Session of Working Group II of the Intergovernmental Panel on Climate Change (IPCC) and Thirtyeighth Session of the IPCC: 25–29 March 2014." *Earth Negotiations Bulletin* 12 (596): 1–20. Accessed May 14, 2014. http://www.iisd.ca/climate/ipcc38/

IPCC WGI. 2013. *Climate change 2013, The physical science basis: Working Group I contribution to the fifth assessment report of the Intergovernmental Panel on Climate Change.* IPCC WGI. Accessed May 14, 2014. http://www.climatechange2013.org/images/report/WG1AR5_ALL_FINAL.pdf

IPCC WGII 2014. *Climate Change 2014, Impacts, Adaptation and Vulnerability.* IPCC WGII. Accessed March 30, 2014. http://ipcc-wg2.gov/AR5/

Ranger, N. 2013. "TOPIC GUIDE: Adaptation: Decision Making under Uncertainty." Evidence on Demand, with DFID. Accessed May 14, 2014. http://www.evidenceondemand.info/evidence-on-demand-ebulletinissue-3-june-2013

World Bank. 2013. *Turn Down the Heat: Climate Extremes, Regional Impacts, and the Case for Resilience.* Washington, DC: The World Bank.

Experiences applying the climate resilience framework: linking theory with practice

Marcus Moench

This paper discusses the evolution and application of the Climate Resilience Framework (CRF). The framework focuses on the roles of systems, agents, institutions, and exposure in climate resilience and adaptation, and supports planning and strategic policy development using iterative shared learning techniques. Conceptual foundations of the CRF are explored, along with its application in a range of implementation and research contexts, including: urban planning (Asia), food systems (Nepal, Central America), and post-flood recovery (Pakistan, USA). These illustrate how analysis of system dynamics and agent behaviour in different institutional contexts can be used to identify points of entry for building resilience.

Cet article traite de l'évolution et de l'application du Cadre de résilience au changement climatique (*Climate Resilience Framework* – CRF). Ce cadre se concentre sur les rôles des systèmes, agents et institutions et sur l'exposition dans la résilience et l'adaptation au changement climatique, et il soutient la planification et l'élaboration de politiques stratégiques au moyen de techniques d'apprentissage itératives communes. Les fondations conceptuelles du CRF sont examinées, ainsi que son application dans une variété de contextes de mise en œuvre et de recherche, y compris : la planification urbaine (Asie), les systèmes alimentaires (Népal, Amérique centrale) et le relèvement post-inondations (Pakistan, États-Unis). Ces contextes illustrent comment l'analyse de la dynamique des systèmes et du comportement des agents dans différents contextes institutionnels peut être utilisée pour identifier des points d'entrées afin de renforcer la résilience.

El presente artículo examina la aplicación y la evolución seguida por el Marco de Resiliencia Climática (CRF). Mediante el uso de técnicas iterativas de aprendizaje compartido, este marco, que opera como un auxiliar para la planificación y el desarrollo de políticas estratégicas, se centra en los roles de los sistemas, los agentes, las instituciones y la experiencia en el ámbito de la resiliencia y de la adaptación al clima. Asimismo, el artículo examina los fundamentos conceptuales del CRF y sus aplicaciones en numerosos contextos de implementación y de investigación, entre los cuales se incluyen la planificación urbana (Asia), los sistemas alimentarios (Nepal, Centroamérica), y la recuperación posinundaciones (Pakistán, EE. UU.). Tales contextos ilustran cómo el análisis de dinámicas de sistema y el comportamiento de los agentes en distintos entornos institucionales pueden ser utilizados para identificar temas de abordaje a través de los cuales ir construyendo la resiliencia.

A framework for understanding climate resilience

Resilience, the ability to bounce back and withstand disruption, is increasingly at the centre of debates over development and responses to climate change. The Resilience Alliance defines the resilience of a social-ecological system (SES) as: *"the ability to absorb disturbances, to be changed and then to re-organise and still have the same identity (retain the same basic structure and ways of functioning)"* (Resilience Alliance 2014). This definition resonates strongly with the challenges of an increasingly interconnected world where climate and other change processes threaten development progress and tensions exist between change and continuity. Like all high-level concepts however, translating resilience into practice requires frameworks that relate basic scientific understanding of SES dynamics to much more specific factors.

The framework developed through work by ISET and numerous other partners for understanding climate resilience has been outlined in recent work through the Asian Cities Climate Change Resilience Network (ACCCRN) (Tyler and Moench 2012; Moench, Tyler, and Lage 2011). Although this focuses on urban areas, the framework was developed on the basis of a much wider array of theoretical and applied research over several decades mostly, though not exclusively, in South Asia.

Elements of the framework emerged initially in debates over integrated water resources management (IWRM). While the importance of integrated understanding in managing any complex system is fully acknowledged, we found that attempts to integrate numerous and often fundamentally different considerations into a single overview perspective often obscured key factors (Moench, Caspari, and Dixit 1999; Moench et al. 2003). In the IWRM case, most applied work focuses on high-level basin management agencies attempting to address competing needs within a river basin. In comparison to prior sector-focused approaches to water management this represented a major advance. In practice, however, IWRM was often framed as a command and control strategy for optimal allocation of available supplies that ignored the diverse array of actors, institutions, and often very localised system characteristics that determine water use and allocation. It involved attempts to reduce variability in ways that ultimately increased system rigidity and the consequences of failure – what Holling and Meffe (1996, 328) call *"the pathology of natural resource management"*. Furthermore, our field research often identified emergent water outcomes that could not be planned based on high-level analysis. They reflected interactions between a wide array of actors and system components and were often full of surprises and unanticipated interactions at local levels and across scales (Moench et al. 2003).

Although initially identified in work on water resources, very similar issues are common in our more recent work on urbanisation, food systems, and disaster risk management. Urban systems are widely recognised as combining a diverse array of interacting human and physical (infrastructure and ecological) elements (Pickett et al. 2001; Alberti et al. 2008). This is also the case with disaster risk (Wisner et al. 2004) and food systems (Ericksen 2007; Ericksen, Ingram, and Liverman 2009). So far, however, attempts to successfully integrate analysis and incorporate results in response processes have had limited success. As a result, broad streams of research on complex systems now emphasise the local contextual factors that contribute to emergent patterns of vulnerability and resilience.

Overall, integrated framing can de-emphasise the role of specific system components, the incentives driving actors, and the formal and informal institutional arrangements that bound behaviour. As a result, attempts to develop fully integrated perspectives can obscure as much as they reveal. Furthermore, attempts at integration often lead to command-and-control or predict-and-prevent recommendations with the expectation that these can be implemented in a top-down manner. Attempts often fail, however, because they de-emphasise the complex factors at play in local contexts and the interests and behavioural drivers of local communities, the private sector, and local government actors. Most attempts at integration are not designed with specific processes to deeply engage actors, understand the nuances inherent in local contexts, and support learning and the inevitable need to adjust course. The focus is often on integration rather than the need for inclusive and adaptive process to build ownership, encourage learning, and support change as conditions evolve.

Given the above, we structured the CRF as both an analytical framework designed to encourage insights across system-agent-institutional boundaries, *and* as an iterative planning process where shared learning would build understanding and encourage adaptive responses (Figure 1). Rather than integration, the goal is to support ownership, recognition, and response as conditions evolve and knowledge grows. It is designed for *emergent* needs rather than integrated prediction and response.

The two circles represent an iterative flow of shared learning through analysis and action to build resilience. The left side describes a vulnerability diagnostic phase, while the right side focuses on the steps that can be taken. The process involves analysis followed by identification of context-specific actions, prioritisation, design, implementation, and monitoring before returning to the basic diagnosis. It was structured to respond to the needs of planning organisations and could be changed for other uses. This paper focuses is on the characteristics of systems, agents, and institutions that contribute to resilience, rather than the process; consequently, it focuses on the diagnostic aspect. It is important to emphasise, however, that *iterative processes* for building understanding, engaging different groups of agents, and adapting approaches are central to resilience and are fundamental to the CRF.

The resilience diagnosis reflects basic distinctions between the characteristics of physical *systems*, *agents*, *institutions*, that determine their resilience when *exposed* to disruption. These distinctions reflect the fact that social-ecological systems (SES) are composed of physical elements, the human individuals and organisations that manage or use them, and the institutional "rules in use" that structure behaviour. Each element reacts differently to stress and the resilience

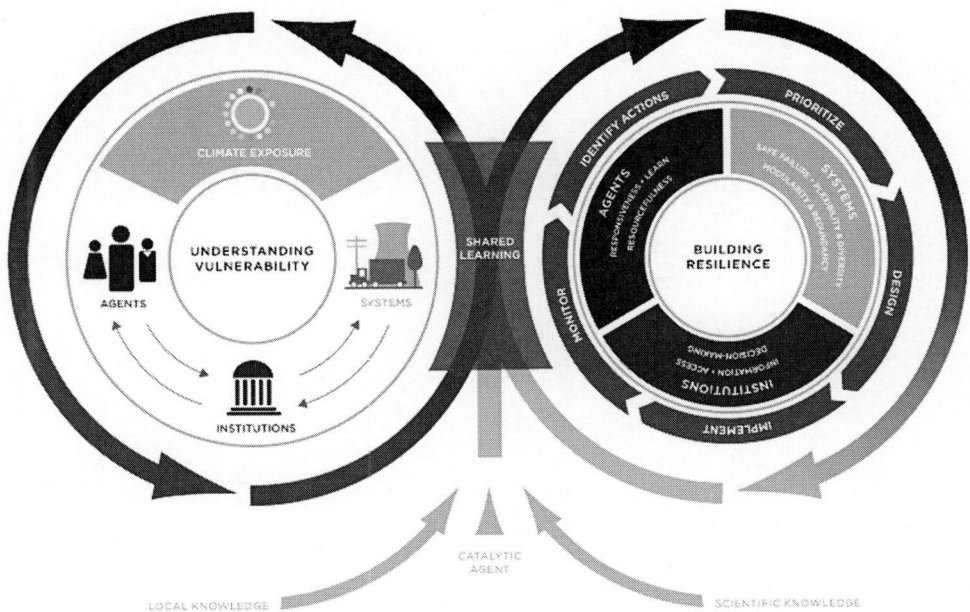

Figure 1.   Climate Resilience Framework.
Source: Friend and MacClune 2012, adapted from Moench, Tyler, and Lage 2011.

of each contributes to resilience of the SES as a whole. Resilience as an outcome – the overall ability of the SES to spring back or evolve – is an emergent property that reflects characteristics of the components and the stresses they are exposed to.

When subject to the sudden-onset types of disruption anticipated as a consequence of climate change, relationships within SES are structured hierarchically (Figure 2). High-level systems, such as markets, social networks, and organisations depend in fundamental ways on underpinning communications, transport, shelter, finance, water supply, and other infrastructure. These systems depend, in turn, on even more basic energy and environmental infrastructure. Without power, communications, transport, and many shelter systems do not function; without communications and transport, organisations, markets, financial systems, and social networks cease to function. As a result, when subject to sharp sudden-onset, hydro-meteorological events, some of the largest potential for cascades is outward from critical ecological and infrastructure systems to the diverse forms of social organisation they support (although the reverse is likely to be true with social forms of disruption such as war). At the same time, however, the ability to respond to system failures, to maintain systems and to adjust as conditions evolve resides in the agents and institutions at higher levels. These higher-level elements are the main source of creativity and adaptive capacity. As a result, resilience of an SES as a whole depends on the characteristics of the core systems, the capacities of agents and organisations at higher levels, and on the institutional features that mediate the interactions between systems and agents.

Systems

Where physical infrastructure or ecosystems are concerned, characteristics that have been widely documented as contributing to resilience include: *flexibility and diversity, redundancy and modularity,* and *safe failure* (Tyler and Moench 2012). These aggregate characteristics reflect a

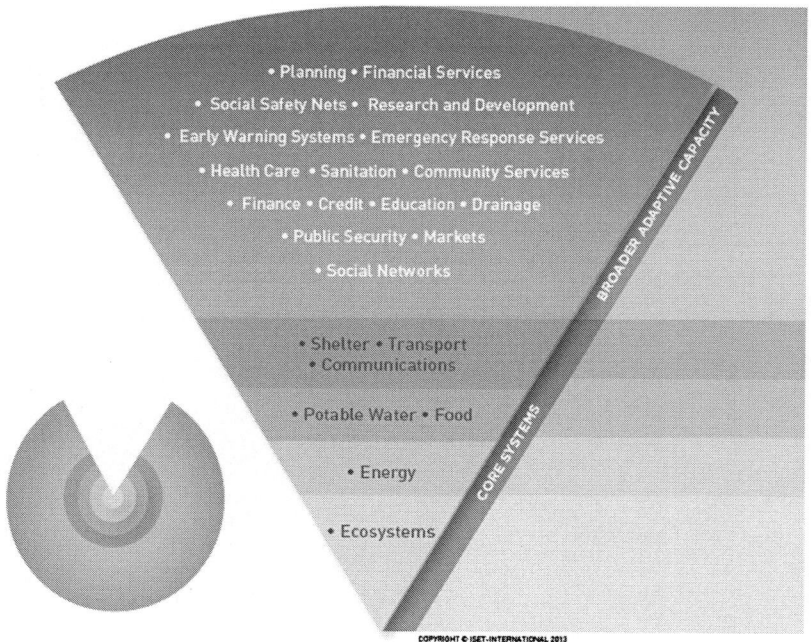

Figure 2. Systems Diagram.
Source: Adapted from Moench, Tyler, and Lage 2011.

number of underlying attributes. Multiple pathways for service provision, the ability to reconfigure key elements to serve multiple functions, locational distribution, and reliance on interchangeable units all contribute (O'Rourke 2007).

These characteristics have practical implications. In a water supply system for example, diverse surface and groundwater supplies, a diversified piping network, and independently powered pumping units reduce the risk that failures will disrupt service provision. At the same time, however, the density of systems and interdependencies between them can undermine resilience even when first-order conditions are met. Water systems, for example, depend heavily on power supply and failures in either system can cascade across both either directly or through physical proximity (O'Rourke 2007). Resilience also often depends on the nature of the disruption. An energy system that, for example, has all the characteristics of flexibility, diversity, redundancy, modularity, and safe failure in relation to hydro-meteorological events may not have those characteristics in relation to fuel availability or economics. As a result, system characteristics must be evaluated in relation to potential sources of disruption. *Resilient to what* is often a key question.

Beyond the characteristics of individual systems, the resilience of regions often depends on the resilience of the underlying support systems. The ability of people and organisations to respond effectively following disaster, for example, depends on whether or not they can access key services. Communication, mobility, power supply, food, and water are basic requirements. Services provided by these systems in turn enable different forms of social organisation and interaction – from hierarchically structured governmental organisations to markets, social networks, and mass movements. When critical systems and the services they provide are disrupted by extreme climatic or other events, this disrupts the functioning of most forms of social organisation. Over longer time periods, however, the reverse is often true. The nature and condition of physical systems depends heavily on how they are managed, maintained, and in some cases designed by individuals and organisations.

The physical factors contributing to resilience in systems are probably better understood than those contributing to agent capacities and shaping institutions. As a result, the following discussion of agents and institutions explores capacities and institutional characteristics in greater depth than the systems characteristics above.

Agents

Where agents are concerned, critical capacities can be summarised as the *ability to learn*, *resourcefulness*, and *responsiveness* (Tyler and Moench 2012). Each characteristic has a range of features that contribute to their resilience function.

Ability to learn

Learning is of fundamental importance if agents are to change strategies as climate evolves. Numerous factors, however, influence the ability of individuals and organisations to learn (Engel and Carlsson 2002; McDaniels and Gregory 2004; Berkhout, Hertin, and Gann 2006). People discount events that are seen as rare, uncertain, or only relevant in the distant future. This is well documented in economics and its appropriateness widely debated in relation to climate change (Sherwood 2007; Heal 2009). They also face difficulty in responding to incremental change processes (Glantz 1999). This implies that the ability to learn depends heavily on the specific manner in which climate evolves. Flexibility and the willingness/opportunity to reflect on experiences are also central to learning (Fazey, Fazey, and Fazey 2005). Organisations often have strong filters related to their cultures, business models, or structures that limit learning. Those founded on particular belief and ideological systems or that limit internal information flow may have particularly strong filters (Hassink and Lagendijk 2001). Overall, while learning is essential, it is not just a matter of training or exposure to issues. Instead, it depends on deeply embedded social factors and how those interact with change processes.

Resourcefulness

Individuals and organisations cannot respond to events if they lack access to basic financial and human capacities. Assets and skills that are convertible as conditions change are of particular importance. Extensive suicides by farmers in drought affected areas of India are, for example, related to financial constraints (debt and assets, such as land, that are poorly convertible) and human capacity limitations (lack of experience with non-farm livelihoods) (Dongre and Deshmukh 2012; Behere and Behere 2008). Other resources, such as social or business networks and livelihood diversification also contribute to resilience by enabling access to assets and capacities.

Responsiveness

Responsiveness to stress from disruptive events is central to resilience. This depends on the depth and frequency of exposure to events (Gunderson and Holling 2002; Tanner et al. 2009); behavioural, belief system, and other forms of lock-in (Scheffer and Westley 2007); and technological or institutional barriers. In addition, responsiveness may depend heavily on psychology. Significant differences exist in recovery from trauma. Large-scale population studies indicate that some individuals return to their prior psychological condition while others are disrupted over the long term or decline after a period or apparent normality (Bonanno, Westphal, and Mancini 2011; Bonanno et al. 2012). This appears to be influenced by personal characteristics and external factors such as finances. It also depends on the psychological ability to filter and interpret information. Many factors that determine responsiveness are, as a result, local. They relate to experiences in daily life and to the social narratives used to filter information and bound perceptions of appropriate responses.

Institutions

Institutions mediate interactions among agents and between agents and systems. Consequently they influence how agent capacities can be deployed and responses governed as stress accumulates. Key institutional features that structure the governance environment include: *rights and entitlements*, *decision-making structures*, *access to information*, and *processes that support learning and change* (Tyler and Moench 2012).

Rights and entitlements

Rights and entitlements determine the ability to access resources and information, to change behaviours or location, and to exercise voice in planning and decision-making processes. The role of entitlements in building social resilience is widely recognised following the work of Dreze, Sen and others on famine and related issues (Dreze, Sen, and Hussain 1995). Entitlements and rights as a result feed directly into the capacities of agents that are, themselves, central to resilience.

Decision-making structures

Resilience is heavily influenced by decision-making structures. Disaster response is almost always initially local. As a result, when systems are disrupted, the authority of local organisations to make decisions is central to resilience. Since risks are shaped by numerous local factors, this is also the case over the longer term. The diverse array of factors involved also represents a strong argument for highly inclusive approaches to decision-making. This said, the long history of natural resource management also highlights frequent conflicts between interests in different areas. Conflicts between upstream and downstream communities in river basins, for example, are ubiquitous. As a result, while decentralisation contributes to the immediate ability to recover from disruption, higher-level checks and balances are also essential.

Access to information

Although rights influence the ability to access information, additional factors are also important. Scientific information on climate change, for example, is often not used due to the absence of collaborative, transparent, and accessible processes for research and communication (Knapp and Trainor 2013). In addition, the technical nature of the subject and inherent uncertainties make detailed understanding essential in order to access the information content of many scientific outputs. As a result, mechanisms for translating information across scales and institutional boundaries are central to access.

Processes that support learning and change

Embedded social narratives and other factors filter information and constrain the ability to learn or change as conditions evolve. As a result, institutional arrangements that support learning and change are central to resilience. Specific mechanisms for iteration and incorporation of new information in strategies are widely recognised as central to this in adaptive management literature. Evidence on how this can consistently be achieved, however, remains limited.

Overall, while elements of the institutional environment that are central to resilience are clear, the characteristics of each element are important. Evidence on consistently effective mechanisms for shaping the institutional environment to produce these characteristics is limited. This is a critical area for additional research. As Olsson and others have argued: "*the institutional and organisational landscapes should be approached as carefully as the ecological in order to clarify features that contribute to the resilience of social-ecological systems*" (Olsson, Folke, and Berkes 2004, 87).

Exposure

Exposure is the final major element central to understanding resilience within the ISET framework. While resilience is often described as independent from the type of stress, this is at best partially true. In the climate change case, for example, social responsiveness depends on relative changes in the frequency of extreme events. In addition, scale is important. Regions will be exposed to both direct localised impacts (such as extreme storms) and indirect impacts through, for example, changes in global food or regional hydrologic systems. Stresses from climate change will translate through the interlinked nature of systems both upward from local events to the global and in the reverse direction. Local events, such as the flooding in Bangkok during 2010, have global implications when they disrupt supply chains. The reverse is also true. This has important implications for thinking about the impacts of climate change. The systemic features (cross-scale diversification) that contribute to resilience to one form of stress can undermine resilience to other forms of stress. There is a balance between the benefits of diversification versus the potential for cascades that depends as much on the nature of the stress as it does on the nature of the system.

In addition to cross-scale issues, the resilience of systems differs according to the nature of stress at the local level. The factors that contribute to the resilience of buildings during earthquakes are, for example, different from those that contribute to flood resilience. Flexibility and safe failure in the earthquake context, for example, is often achieved by the use of materials (such as wood and drywall) that are subject to immediate damage by flooding, whereas materials that are flood resilient (brick and cement) are difficult to utilise safely in earthquake prone areas. Similarly the ability to respond and recover from sudden onset events is often enhanced by very decentralised decision-making processes, while the ability to respond to slower, more incremental, or cross-scale events require countervailing checks and balances on the decision-making power of local interests. Overall, while basic resilience principles apply broadly, outcomes depend heavily on the nature of the disruption. While it is impossible to fully predict patterns of climate change, building resilience also requires understanding the types of stress that are likely.

Domains of resilience

The characteristics of agents, institutions, and systems that contribute to resilience are likely to be most effective when they work in synergy rather than isolation. Conceptually, the characteristics aggregate in complementary and reinforcing ways. As with pieces of a mosaic, each characteristic is particularly meaningful in conjunction with other characteristics. When underlying systems are resilient, they reduce the level of disruption from any given event while at the same time increasing the ability of agents to respond. Similarly, when institutions enable access to information and support learning, people can identify emerging stresses and respond. As a result, where the contributing characteristics are present across a broad range of systems, agents, and institutions *domains of resilience* are likely to emerge. Urban areas may often display high levels of resilience because the density of different systems and agent capacities combined with the fluidity of institutions generates many of the relevant attributes. This may also be the case with development processes more generally.

While development often proceeds in ways that are vulnerable to disruption, the process often results in the creation of highly redundant and layered systems, contributes to agent capacities, and breaks down numerous constraining institutions. As a result, while development and resilience are not equivalent, the two can be complementary – much depends on whether or not resilience characteristics are actively incorporated or indirectly generated by the development process. The reverse is also true: development processes that result in the creation of fragile systems undermine resilience. This may underpin the vulnerability of countries in locations such as South Asia and Africa. Resilience in these areas is low because systems are fragile, most agents have limited capacities, and institutional constraints are widespread. The net result is a domain of vulnerability that contrasts strongly with the domains of resilience in regions where key characteristics exist across a broad spectrum of systems, agents, and institutions.

A final point is that the resilience of a state is not necessarily indicative of its desirability. Exploitative systems can be extremely resilient: for example, systems composed of redundant, flexible, and highly replaceable units – such as large populations of poor workers – can be highly resilient. Similarly, institutions that contribute to the capacity of one class of agents build the resilience of *those agents* and the social structure that benefits them. The narrative of resilience as a positive characteristic must, as a result, recognise that resilience is *in itself* a neutral attribute. Resilience concepts address the ability of social-ecological systems to *"absorb disturbances, to be changed and then to re-organise and still have the same identity"* (Resilience Alliance 2014). Taken alone these concepts do not address resilience *of whom, by whom, or for what purposes* (Friend and Moench 2013).

Applications in resilience research and planning

Although elements of the CRF were developed through numerous projects, it crystalised as an integrated framework during the course of the Asian Cities Climate Change Resilience Network (ACCCRN) programmes, supported by the Rockefeller Foundation. ACCCRN was developed to provide support to cities across India, Vietnam, Thailand, and Indonesia in building climate resilience. Approaches developed in ACCCRN have been refined and implementation expanded through the USAID-supported Mekong–Building Climate Resilience in Asian Cities (MBRACE) programme work in additional cities across Vietnam and Thailand. Beyond this, ISET has used the integrated framework for research in a wide variety of contexts primarily in Asia but also in Latin America and our home location (Boulder, Colorado) in the USA. Here, we draw on examples from ACCCRN; local adaptation planning and analysis of food systems in Nepal; work on food system resilience indicators with the International Institute of Sustainable Development (IISD) in Central America; post-flood recovery in Pakistan; work on the economics of climate resilient shelter for the Climate Development Knowledge Network (CDKN); and analysis of the September 2013 floods in Boulder, Colorado, for the American Red Cross.

A detailed discussion of the ACCCRN programme, including case studies and in-depth discussion of the CRF, is available elsewhere (Moench, Tyler, and Lage 2011; Tyler and Moench 2012; Friend and Moench 2013). Reports for the work on food systems and adaptation planning in Nepal, research on post-flood recovery in Pakistan (Khan 2013), and aspects of work under the MBRACE programme are also available (ISET, NISTPASS, TEI 2012; ISET, TEI, NISTPASS 2013). Analysis of food systems in central Nepal using the CRF was undertaken by ISET-International and ISET-Nepal during 2010–2012 with support from the Canadian International Development Research Centre (IDRC). Both also worked with WFP to develop resilience indicators for Nepal's food security monitoring system with support CDKN (Dixit et al. 2013). Work is ongoing on the CDKN-supported projects on *Sheltering from a Gathering Storm*, and with IISD on *Climate Resilience and Food Security in Central America* (CREFSCA). This is also the case with the study of flooding in Boulder, Colorado, supported by the American Red Cross. Publications from these projects are anticipated in 2014.

Although the applications differ, in each of the above cases core elements of the framework were central. All have: (1) used the interaction between systems, agents, institutions, and exposure as a basis for diagnosing vulnerability; (2) incorporated shared learning methods; and (3) used results to identify points of entry for potential interventions. In addition, the basic approach used in all of the cases (with the exception of the ongoing research in Boulder) was the same. It involved a shared learning process that proceeded from initial engagement through analysis to the identification of measures for monitoring or potential courses of action to build resilience. In most cases the process actively engaged local communities, government agencies, and other organisations in diagnosing vulnerabilities and identifying courses of action. In ACCCRN and MBRACE, this was taken one step further and involved the development of strategic resilience plans. These strategic plans were intended as a basis for consolidating, prioritising, and implementing resilience measures.

Despite differences in context, the climate concerns that emerged were very similar. Climate change as such was rarely the central concern. Data on climate were difficult to obtain, interpret, or use. It became clear that discussions on climate rarely drill down to the level where stakeholders are active. Discussions tended, as a result, to focus on immediately tangible issues that could be related to climate change such as urban planning, flooding, water supply, drainage, and storm impacts. Second, discussions tended to move quickly from broad issues to specific short-term activities. What emerged was a pattern of discussion that focused on *who* could take *what* action in relation to *which* general climate concern, and its impacts on both *what* (physical or social component) and *which* group. Groups also tended to separate systems into environmental and infrastructure categories. This reflected the fact that environmental and infrastructure issues are managed by different organisations with different skill and interest sets. Reflecting this, discussions emphasised the realities of who could (and had an incentive to) address specific concerns. Areas of potential activity tended to be scaled to reflect the perceived financial and other capacities of local-level actors and the external resources available. Aggregate perspectives on climate, urban systems, or regional issues, while useful for context, provided little insight into potential courses of action.

These highly diverse contexts illustrate how the CRF as a unified framework for analysis assists local actors in understanding complex issues and identifying very practical points of entry for building resilience. Furthermore, it helps to locate individual activities within a wider, more complete, understanding of how they contribute to climate resilience. This is important because many activities that contribute to resilience are simply good development and risk management practice. In isolation, it is often difficult to identify their unique contribution to "climate resilience" as such. Instead, it is the context and combination of interventions within the larger strategic understanding provided by the CRF that clarifies the contribution any activity makes toward larger resilience objectives.

The types of activity that were undertaken in each of the projects and how they relate to the main elements of the CRF are summarised in Table 1.

In isolation, most of the activities identified in Table 1 could be seen as standard environment and development or disaster risk management research and implementation activities. This is to be expected. The difference, however, lies in the aspects of systems, agents, and institutions that when aggregated relate in an overall sense to climate resilience. Beyond the activities themselves, several points are important to note.

First, the interests of agents and the institutional context in which they operate were central in the identification of potential activities. For any activities to move forward, the process had to engage with different sets of agents, recognise their scope of action, and respond to the priorities they saw as important. While the impacts of climate change were often discussed, this element of agency (*who* might do *what* and *why*) is rarely addressed explicitly in discussions around climate change.

Second, the types of activities that emerged in each case fit well with the basic underlying characteristics of resilience synthesised in the CRF. Despite the broad engagement process and diverse set of contexts, the basic systems-agents-institutions-exposure structure and underlying attributes could be applied effectively in all cases.

Third, few, if any, activities could be seen as relating to resilience on their own. Instead, they could best be characterised as strategically selected elements contributing to a larger mosaic of characteristics.

Strengths and limitations of the CRF

Where strengths are concerned, the framework and approach have drawn attention to new insights and helped build capacities and institutions. The ACCCRN programme has catalysed awareness and activity in the cities it has engaged with on climate resilience, helped develop institutions, built capacity, and resulted in the implementation of a broad base of activities. Analysis using the CRF in Nepal and Central America highlights numerous considerations that affect food security but are not captured in more conventional locally focused analyses. Historically, most analyses of food security in Nepal and Central America have emphasised localised production-consumption-access issues. Use of the CRF drew attention to the changing nature of food systems and emphasised the growing role of non-farm livelihoods, remittances, and regionally concentrated

Table 1.   Mapping of project activities to core elements of the CRF.

Attributes Addressed	Programmes					
	ACCCRN and MBRACE	CDKN Sheltering	Nepal Food Security	Latin America Food Security	Pakistan Floods	Boulder Floods
Agents						
Responsiveness	Improve public health surveillance, disaster response, and access to early warning information	Capture and advertise stories of benefits following events	Indicators developed for WFP and Nepal Government food-security monitoring system		Systematically explore the factors influencing the ability of HH to recover from floods	Extensive interviews on how agencies and home owners responded both through official and autonomous action during the floods
Resourcefulness	Promote alternative livelihoods and implement credit programmes for poor. Include multiple sectors in planning and build understanding of capacities and knowledge.	Credit programmes for resilient household structure development, cost-benefit analysis	Livelihood and food source diversification	Help communities identify new areas where they can take action to improve food security	Develop evidence base to improve access to international finance for system resilience	Research factors influencing the ability of individuals, private sector and government to access finance, personnel, equipment, and information during and after the flood
Learning	Incorporate climate change into school curriculum, training in CBDRM and disaster response; build monitoring mechanisms in climate planning and intervention projects; institutionalise SLDs. Build reflection and learning into city processes and into programme M&E	Increase awareness of design elements among masons, builders, and public	Engage communities in analysis of climate vulnerability through Shared Learning process	Engage communities and NGOs in process based on CRF for understanding food security; application of SLDs	Engage government and other actors in research process	Hold interviews and discussions on local public radio, assess the ability of local organisations to draw lessons from the floods, and participate in long-term recovery planning

(*Continued*)

Table 1. Continued.

Attributes Addressed	Programmes					
	ACCCRN and MBRACE	CDKN Sheltering	Nepal Food Security	Latin America Food Security	Pakistan Floods	Boulder Floods
Systems **Infrastructure** Safe-failure	Hydrologic and hydraulic modelling for infrastructure design, improve early warning systems. Enhance systems thinking through scenario planning visualisations	Shelter designs where damage is incremental rather than catastrophic	Investigate food storage at HH level, investigate transport system alternatives in case of failure in mountain roads	Identify fragilities in food system infrastructure (transport and cooling) where failures cause major disruption	Research to identify where system failures (flood control, transport, sanitation, water supply) have minimal or major impacts and the design factors that contribute	Investigate factors contributing to safe- and unsafe failure of water, irrigation, sewage and transport infrastructure. Identify factors that caused failure (same factors as in Asia). Identify areas that could benefit from learning and autonomous action.
Flexibility and diversity	Storm and flood resilient housing; diversify municipal water supply sources	Diverse designs for diverse conditions	Diversification of access to markets	Discuss potential avenues for diversifying transport and cold storage	Research diversity within shelter and other systems for application in different flood contexts	Identify options for diversifying power and water supply access. Identify and highlight where flexibility and diversity built unexpected resilience
Redundancy and modularity	Storm and flood resilient housing; community based drainage and canal restoration	Housing highly modular	Investigate access to key crops (rice) from local, regional and global sources			Identify options for diversifying sewage treatment and/or preventing sewage backup damage, bolstering multiple options for the potable water system, and analysing where the transport system broke down

Environmental

Safe-failure	Watershed planning and forest protection, Mangrove restoration	Shelter designs effective where environmental conditions moderate levels and flow strengths	Focus on food types that can substitute for each other (roles of different grains) if one type fails		Explore role of flood buffer areas (wetlands) in mitigating potential for major infrastructure failure	Investigate effective functioning of urban floodplain management in city
Flexibility and diversity	Mangrove as well as structure; biostructural riverbank stabilisation		Possibilities for diversification of crops	Identify options for diversifying crops; home gardens	Explore diverse needs for sanitation depending on environmental conditions (arid zone, inundation areas, etc.)	Highlight need for diversity in responses and management based on environmental zones
Redundancy and modularity	Groundwater recharge to supplement surface availability for water supply		Role of different ecological zones in producing similar crops	Identify options to increase redundancy and diversity in irrigation systems		
Institutions Rights and entitlements including access to information	Improve climate warning and forecasting services to diverse users	Host shelter design competitions with extensive publicity	Early warning on food system fragility through indicators	Improve access to information on food systems; access to supporting health services and education		Identify where current practices and policies limit the ability of organisations and government to support autonomous response, and where government/civil society interactions manage the difficulties of assisted undocumented immigrants

(Continued)

Table 1. Continued.

Attributes Addressed	Programmes					
	ACCCRN and MBRACE	CDKN Sheltering	Nepal Food Security	Latin America Food Security	Pakistan Floods	Boulder Floods
Decision-making structures	Water demand management, limit development rights in flood zones, establish climate coordination offices	Focus on builders and home owners as decision makers		Identify needs to include government and other stakeholders in food system analysis; recognition of local governance as a vulnerability factor	Review existing policies on flood management and decision-making	Highlight limitations in codes, floodplain zoning and other regulatory structures, and the ability of all stakeholders to participate in the recovery process
Processes supporting learning and change	New coordination and technical support organisations; engage communities in resilience planning; climate resilience planning processes	Shared learning and community engagement processes around shelter; resilient shelter design competitions	Develop shared learning processes for local adaptation planning	Develop indicators and shared learning processes for communities	Implement shared learning process with government and communities	Support emerging resilience strategy development initiatives and long-term recovery planning. Propose community learning and engagement processes

Exposure						
Current	Historical frequency of extreme flood and storm events, existing infrastructure and development patterns in exposed locations	Collection of statistical information on historical events and losses for storms and floods.	Synthesis of available historical information on food system failures related to weather	Synthesis of available historical information on food system failures related to weather	Collection of historical information on major flood events with particular focus on 2010. Identify how settlement patterns and institutional decisions influence exposure	Review of historical information on flooding, fire and other extreme events. Interviews with key system managers.
Future	Downscaling and other analyses of emerging scientific information on climate. Local development processes and expansion into exposed areas	Use CCSM4 projections and local data on temperature and precipitation (where possible). Literature on changes in storm characteristics and analysis of major development trends	Compilation of available scientific information on climate change. Emphasis on high uncertainties in Himalayan region.	Compilation of available scientific information on climate change. Emphasis on uncertainties in region.	Review of global literature on implications of climate change for flooding in the region. No additional modelling.	Engagement with wider working groups on climate in Boulder (very diversified scientific basis).

production and distribution systems. It also suggests points of entry for strengthening food security at different levels (within, for example, distribution systems) and in arenas (such as transport, power systems, and non-farm livelihoods) that have not historically been central to food policy.

Similar findings emerged in the Pakistan flood study, where results clearly showed the correlation between access and duration of usage of improved systems and post-disaster recovery rates. Analysis largely validates the hypothesis that access to basic system services improves resilience in flood-affected areas (Khan 2013). It also, however, makes clear that access to improved systems may not be the best measure of system quality. Where traditional or other mechanisms provide reliable access to core services (such as clean drinking water), access to improved systems does not aid recovery. This highlights the importance of analysing system characteristics and access locally to determine their contribution to resilience and disaster recovery. The relevance of any specific type of system is situational. Where the Boulder flood case study for the ARC is concerned, use of the CRF has provided the team involved with both a clear structure for identifying key areas to focus research on and a readily understood basis for dialogue with diverse groups of stakeholders. This activity is at an early stage but is important because it clearly illustrates the applicability of the CRF in high-capacity wealthier regions as well as less industrialised areas.

Limitations of the CRF as an analytical and approach framework are also substantial. First, the activities that have been identified in each of the programmes shown in Table 1 are incremental elements in what ultimately needs to become a larger mosaic of characteristics. The CRF and associated shared learning process help in identifying strategic points of entry for building resilience. They also assist in prioritising activities. They cannot, however, address the inherent limitations of existing institutions, the amounts of resources required to address system fragilities, or the scale at which local agents are able to act. Second, use of the CRF tends to guide thinking toward planning and activities through governmental organisations rather than autonomous adaptation incentives and behaviours (the agency of local populations). Third, as with most process-driven approaches, the nature of the insights generated and activities identified depend heavily on the degree and diversity of stakeholder involvement. Use of the CRF does not ensure that the issues of most importance to the poor or specific vulnerable groups will emerge *unless* they have a strong presence and voice in the shared learning processes.

Conclusions

The CRF provides a structure for vulnerability analysis that combines systemic, behavioural, and institutional factors. This enables insights into the sources of resilience and vulnerability that are different from those generated through either human-centred approaches or those based on systems analysis alone. By disaggregating approaches to analysis and drawing attention to the interactions between the behavioural drivers of agents, institutional contexts, system dynamics, and changing patterns of exposure, it enables identification of *targeted* points of entry at different scales where interventions can build resilience. These relate to characteristics (such as diversification, flexibility, the ability to learn, etc.) that are fundamental to both human and systems resilience. In conjunction with shared-learning and iterative planning, this creates an adaptive mechanism for translating insights into practical courses of action over time. For the framework to be useful, however, considerable skill is required in adapting its use to different contexts. Substantial translation and interpretation is required for use with practitioners or policy actors. Finally, work is required to clarify how many of the characteristics that contribute to resilience are hazard-, culture-, or otherwise specific. While evidence suggests that the basic characteristics of systems, agents, and institutions that contribute to resilience are generalisable, such frameworks cannot be applied blindly.

Acknowledgements

The author would specifically like to acknowledge inputs from the following individuals outside ISET-International: Ajaya Dixit (ISET-Nepal), Dipak Gyawali (Nepal Water Conservation Foundation), S. Jankaarajan (MIDS), Simon Anderson (IIED), Jo D'Silva (Arup), Anil Pokhrel (currently World Bank), and Shiraz Wajhi (GEAG).

References

Alberti, M., J. M. Marzluff, E. Shulenberger, G. Bradley, C. Ryan, and C. Zumbrunnen. 2008. "Integrating Humans into Ecology: Opportunities and Challenges for Studying Urban Ecosystems." In *Urban Ecology*, edited by J. M. Marzluff, E. Shulenberger, W. Endlicher, M. Alberti, G. Bradley, C. Ryan, U. Simon, and C. Zumbrunnen, 143–158. New York: Springer.

Behere, P. B., and A. P. Behere. 2008. "Farmers' Suicide in Vidarbha Region of Maharashtra State: A Myth or Reality?." *Indian Journal of Psychiatry* 50 (2): 124–127. doi: 10.4103/0019-5545.42401.

Berkhout, F., J. Hertin, and D. M. Gann. 2006. "Learning to Adapt: Organisational Adaptation to Climate Change Impacts." *Climatic Change* 78 (1): 135–156. doi: 10.1007/s10584-006-9089-3.

Bonanno, G. A., M. Westphal, and A. D. Mancini. 2011. "Resilience to Loss and Potential Trauma." *Annual Review of Clinical Psychology* 7: 511–535. doi: 10.1146/annurev-clinpsy-032210-104526.

Bonanno, G. A., P. Kennedy, I. Galatzer-Levy, P. Lude, and M. Elfstrom. 2012. "Trajectories of Reilience, Depression, and Anxiety Following Spinal Cord Injury." *Rehabilitation Psychology* 57 (3): 236–247. doi: 10.1037/a0029256.

Dixit, A., Y. Subedi, T. McMahon, and M. Moench. 2013. *Mainstreaming Climate Sensitive Indicators into an Existing Food Monitoring System: Climate Change and Food Security in Nepal*. Kathmandu: Institute for Social and Environmental Transition–Nepal.

Dongre, A. R., and P. R. Deshmukh. 2012. "Farmers' Suicides in the Vidarbha Region of Maharashtra, India: A Qualitative Exploration of their Causes." *Journal of Injury and Violence Research* 4 (1): 2–6. doi: 10. 5249/jivr.v4i1.68.

Dreze, J., A. Sen, and A. Hussain. 1995. *The Political Economy of Hunger*. New Delhi: Oxford University Press.

Engel, P. G. H., and C. Carlsson. 2002. "Enhancing Learning Through Evaluation: Approaches, Dilemmas and Some Possible Ways Forward." Paper presented at *EES Conference*, Seville, Spain, October 10–12.

Ericksen, P. J. 2007. "Conceptualizing Food Systems for Global Environmental Change Research." *Global Environmental Change* 18 (1): 234–245. doi: 10.1016/j.gloenvcha.2007.09.002.

Ericksen, P. J., J. S. I. Ingram, and D. M. Liverman. 2009. "Food Security and Global Environmental Change: Emerging Challenges." *Environmental Science & Policy* 12 (4): 373–377. doi: 10.1016/j. envsci.2009.04.007.

Fazey, I., J. Fazey, and D. Fazey. 2005. "Learning More Effectively from Experience." *Ecology and Society* 10 (2): 1–22. http://www.ecologyandsociety.org/vol10/iss2/art4/

Friend, R., and K. MacClune. 2012. *Climate Resilience Framework: Putting Resilience Into Practice*. Boulder, CO: Institute for Social and Environmental Transition–International.

Friend, R., and M. Moench. 2013. "What is the Purpose of Urban Climate Resilience? Implications for Addressing Poverty and Vulnerability." *Urban Climate* 6: 98–113. doi: 10.1016/j.uclim.2013.09.002.

Glantz, M. H. 1999. *Creeping Environmental Problems and Sustainable Development in the Aral Sea Basin*. Cambridge: Cambridge University Press.

Gunderson, L. H., and C. S. Holling. 2002. *Panarchy: Understanding Transformations in Human and Natural Systems*. Washington, DC: Island Press.

Hassink, R., and A. Lagendijk. 2001. "The Dilemmas of Interregional Institutional Learning." *Environment and Planning C: Government and Policy* 19 (1): 65–84. doi: 10.1068/c9943.

Heal, G. 2009. "The Economics of Climate Change: A Post-Stern Perspective." *Climate Change* 96 (3): 275–297. doi: 10.1007/s10584-009-9641-z.

Holling, C. S., and G. K. Meffe. 1996. "Command and Control and the Pathology of Natural Resource Management." *Conservation Biology* 10 (2): 328–337. doi: 10.1046/j.1523-1739.1996.10020328.x.

ISET, NISTPASS, and TEI. 2012. *Changing Cities and Changing Climate: Insights from Shared Learning Dialogues in Thailand and Vietnam*. Boulder, CO: Institute for Social and Environmental Transition–International.

ISET, TEI, and NISTPASS. 2013. *Assessing City Resilience: Lessons from Using the UNISDR Local Government Self-Assessment Tool in Thailand and Vietnam*. Boulder, CO: Institute for Social and Environmental Transition–International.

Khan, F. 2013. *Indus Floods Research Project*. Boulder: Institute for Social and Environmental Transition.

Knapp, C. N., and S. F. Trainor. 2013. "Adapting Science to a Warming World." *Global Environmental Change* 23 (5): 1296–1306. doi: 10.1016/j.gloenvcha.2013.07.007.

McDaniels, T., and R. Gregory. 2004. "Learning as an Objective within a Structured Risk Management Decision Process." *Environmental Science and Technology* 38 (7): 1921–1926. doi: 10.1021/es0264246.

Moench, M., E. Caspari, and A. Dixit. 1999. *Rethinking the Mosaic: Investigations into Local Water Management*. Kathmandu: Nepal Water Conservation Foundation and the Institute for Social and Environmental Transition.

Moench, M., A. Dixit, S. Janakarajan, M. S. Rathore, and S. Mudrakartha. 2003. *The Fluid Mosaic: Water Governance in the Context of Variability, Uncertainty and Change*. Kathmandu: Nepal Water Conservation Foundation and the Institute for Social and Environmental Transition.

Moench, M., S. Tyler, and J. Lage. 2011. *Catalyzing Urban Climate Resilience: Applying Resilience Concepts to Planning Practice in the ACCCRN Program (2009–2011)*. Boulder: The Institute for Social and Environmental Transition–International.

O'Rourke, T. D. 2007. "Critical Infrastructure, Interdependencies, and Resilience." *The Bridge: Linking Engineering and Society* 37 (1): 22–9. http://www.nae.edu/Publications/Bridge/Engineeringforthe ThreatofNaturalDisasters/CriticalInfrastructureInterdependenciesandResilience.aspx

Olsson, P., C. Folke, and F. Berkes. 2004. "Adaptive Comanagement for Building Resilience in Social-Ecological Systems." *Environmental Management* 34 (1): 75–90. doi: 10.1007/s00267-003-0101-7.

Pickett, S. T. A., M. L. Cadenasso, J. M. Grove, C. H. Nilon, R. V. Pouyat, W. C. Zipperer, and R. Costanza. 2001. "Urban Ecological Systems: Linking Terrestrial Ecological, Physical, and Socioeconomic Components of Metropolitan Areas." *Annual Review of Ecology and Systematics* 32: 127–157. doi: 10.1146/annurev.ecolsys.32.081501.114012.

Resilience Alliance. 2014. "Key Concepts." Accessed May 9, 2014. http://www.resalliance.org/index.php/ key_concepts

Scheffer, M., and F. Westley. 2007. "The Evolutionary Basis of Rigidity: Locks in Cells, Minds, and Society." *Ecology and Society* 12 (2): 1–13. http://www.ecologyandsociety.org/vol12/iss2/art36/

Sherwood, S. 2007. "Discounting and Uncertainty: A Non-Economist's View." *Climatic Change* 80 (3–4): 205–212. doi: 10.1007/s10584-006-9164-9.

Tanner, T., T. Mitchell, E. Polack, and B. Guenther. 2009. "Urban Governance for Adaptation: Assessing Climate Change Resilience in Ten Asian Cities." *IDS Working Papers,* 2009: 1–7. doi: 10.1111/j. 2040-0209.2009.00315_2.x.

Tyler, S., and M. Moench. 2012. "A Framework for Urban Climate Resilience." *Climate and Development* 4 (4): 311–326. doi: 10.1080/17565529.2012.745389.

Wisner, B., P. Blaikie, T. Cannon, and I. Davis. 2004. *At Risk: Natural Hazards, People's Vulnerability and Disasters*. 2nd ed. London and New York: Routledge.

A changing climate for international development

Tim Wheeler

Leave no one behind. That is the key message for reducing poverty worldwide from the recent report, "A New Global Partnership: Eradicate Poverty and Transform Economies through Sustainable Development", from the High-level Panel of Eminent Persons on the post-2015 Development Agenda (2013). It is a bold aspiration, and one that lays down a grand challenge to all those involved in some way with international development and poverty reduction. Many difficult and complex factors need to be addressed to achieve this goal; foremost amongst these is climate change.

The collection of papers in this special journal issue explore many aspects of climate's impact on low-income countries and the lives of the poor. They add to a body of climate science and climate-related development research that has gradually increased in volume and extent since the publication of the first assessment report of the Intergovernmental Panel on Climate Change (IPCC) in 1990. The confidence of the research community that global climate is changing, and the attribution of these changes to human activity, has strengthened dramatically over the last 25 years. And knowledge is steadily increasing on how a changing climate impacts on the lives of the poor.

There is a near consensus of scientific opinion that the globe is warming. Several studies demonstrate clearly that there has been a progressive increase in global mean temperature since the middle of the nineteenth century (Hansen et al. 2010). The scientific consensus expressed in the fifth assessment report of the IPCC of 2013 now attributes, with more than 95% certainty, observed climate change since the mid twentieth century to human activities such as the emission of greenhouse gases (IPCC 2013). Climate change due to human activities is expected to bring warmer temperatures, changes to rainfall patterns, and make extreme weather events more frequent. By the end of this century it is thought that global mean temperature could be up to 4.8°C warmer than at the end of the last one.

There will always be uncertainties associated with trying to project the climate conditions of the world of 20–50 years in the future and their impacts on people and their environment. Nevertheless, it is widely accepted that climate change will have its greatest impact in the developing world, and on the poorest within those communities. This is in part because they tend to live in climates that are already extreme in some way (e.g., seasonally arid) or they are currently vulnerable to the effects of weather extremes (e.g., by living close to sea-level in regions at risk of flooding). These communities are also less able to adapt their lives to reduce the harmful impacts of climate change. Hence, climate change threatens the development of individuals, communities, and nations. A recent assessment of the impacts on global food security, for example, concluded that climate change in developing nations is likely to hinder progress towards a world without hunger (Wheeler and von Braun 2013).

Development challenges from climate will worsen over time as the degree of climate change increases. Most immediate impacts (in the next 10 years or so), however, will be felt through the effects of extreme weather, such as droughts, heat waves, and flooding. Although particular extreme weather events (such as the 2010 floods in Pakistan) cannot be attributed to human influences on the climate, there will very probably be more of them in the future. People who are vulnerable to the effects of extreme weather now, will be even more vulnerable in the future. As we move through this century, what are currently extreme weather events are likely to become seen as normal.

We are already committed to some climate change for the next 20–30 years due to past emissions of greenhouse gases. This means that there is a need to adapt now to the impacts of changes in climate that the world is already committed to. And of course, global greenhouse emissions continue year on year, with over 30 billion tonnes of CO_2 equivalent emitted to the atmosphere in 2012, locking in further climate change over the coming decades.

The evidence base on climate change is now clear on three important issues: that the climate is changing; that it is increasingly certain that climate is changing due to human activities, through the emissions of greenhouse gases and land-use change; and that the poorest people in the world will be disproportionately affected by future climate change. It is surprising, therefore, that current global investments in building better climate adaptation and resilience as part of development assistance programmes is such a low proportion of total development finance (Ranger, Harvey, and Garbett-Shiels 2014). We can be confident that achieving the goal of ending global poverty will be almost insurmountable without a more sustained effort to counter the impacts of climate change.

References

Hansen, J., R. Ruedy, M. Sato, and K. Lo. 2010. "Global Surface Temperature Change." *Reviews of Geophysics* 48: RG4004.

High-level Panel of Eminent Persons on the post-2015 Development Agenda. 2013. "A New Global Partnership: Eradicate Poverty and Transform Economies through Sustainable Development." Accessed February 10, 2014. http://www.post2015hlp.org/the-report/

IPCC. 2013. "Summary for Policymakers." In *Climate Change 2013: The Physical Science Basis. Contribution of Working Group I to the Fifth Assessment Report of the Intergovernmental Panel on Climate Change*, edited by T. F. Stocker, D. Qin, G.-K. Plattner, M. Tignor, S. K. Allen, J. Boschung, A. Nauels, Y. Xia, V. Bex and P. M. Midgley. Cambridge, UK: Cambridge University Press.

Ranger, N., A. Harvey, and S-L. Garbett-Shiels. 2014. "Safeguarding Development Aid Against Climate Change: Evaluating Progress and Identifying Best Practice." *Development in Practice* 24 (4).

Wheeler, T., and J. von Braun. 2013. "Climate Change Impacts on Global Food Security." *Science* 341 (6145): 508–513.

Safeguarding development aid against climate change: evaluating progress and identifying best practice

Nicola Ranger, Alex Harvey and Su-Lin Garbett-Shiels

Official development assistance currently totals around US$130 billion per year, an order of magnitude greater than international climate finance. To safeguard development progress and secure the long-term effectiveness of these investments, projects must be designed to be resilient to climate change. This article reviews 250 projects for three countries from two development organisations and finds that between 2% and 30% of these may require action now to "future-proof" investments and policies. Both organisations show improvements in the recognition of climate change in projects, but many projects are still not future-proof.

L'aide officielle au développement s'élève actuellement à environ 130 milliards de dollars par an, montant supérieur à celui du financement international de la lutte contre le changement climatique. Pour protéger les progrès du développement et garantir l'efficacité à long terme de ces investissements, les projets doivent être conçus de manière à être résilients face au changement climatique. Cet article examine 250 projets menés dans trois pays par deux organisations de développement et constate qu'entre 2 et 30 % d'entre eux requièrent une action dès à présent pour protéger les investissements et les politiques des aléas futurs éventuels. Les deux organisations affichent des progrès sur le plan de la reconnaissance du changement climatique dans les projets, mais nombre de projets ne sont pas encore à l'épreuve du temps.

La ayuda oficial destinada al desarrollo alcanza alrededor de 130 mil millones de dólares por año, un orden de magnitud mayor que el orientado a responder a los efectos del cambio climático a nivel mundial. Con el fin de salvaguardar el avance del desarrollo y de garantizar la efectividad a largo plazo de estas inversiones, los proyectos deberán tomar en cuenta los efectos producidos por el cambio climático. El presente artículo informa sobre los resultados de una revisión efectuada a 250 proyectos impulsados por dos organizaciones de desarrollo en tres países. Se encontró que entre 2 % y 30 % de dichos proyectos requieren actualmente la implementación de cambios que los hagan pertinentes para las inversiones y las políticas a futuro. Ambas organizaciones han instrumentado modificaciones en algunos proyectos de manera que respondan al cambio climático; sin embargo, otros proyectos permanecen sin considerar esta proyección a futuro.

Introduction

Tackling climate change is widely recognised as crucial to achieving long-term sustainable poverty alleviation. Poverty alleviation and climate change are intimately linked (Stern 2007) as the poorest people tend to suffer the greatest impacts and have the least capacity to adapt. Even today climate shocks, like droughts, flooding, and storms, have a material impact on the development prospects of the poorest countries. Since 1980, weather-related catastrophes have caused almost 1.2 million fatalities and led to direct damages amounting to US$610 billion in low income (LICs) and lower-middle income countries (LMICs).[1] For the poorest in society, these direct impacts can have a long-term influence on economic prospects.

Climate change is expected to increase the intensity and frequency of climate shocks in many regions (IPCC 2012). In parallel, gradual changes in climate such as rising temperatures, changing rainfall patterns, and sea level rise, will affect human health, food systems, water supplies, and ecosystems. This will create a more challenging environment for development (World Bank 2010, 2013a). Climate change will also interact with other pressures, such as population growth, urbanisation, resource scarcity, and conflict, which will multiply risks to development.

Without appropriate interventions, climate change could create a vicious circle of growing vulnerability and impacts; the poor could be driven deeper into poverty and the gains achieved through development cooperation may be reversed (World Bank 2010). This risk is high on the political agenda. The May 2013 report of the High Level Panel on the Post-2015 Development Agenda reiterated that *"without tackling climate change, we will not succeed in eradicating extreme poverty"* (UN 2013, p. 55).

Many development agencies acknowledge that adapting to climate change is critical to achieving broader development goals.[2] As part of the Copenhagen Accord, developed countries agreed to mobilise US$100 billion per year to support adaptation and mitigation by 2020.[3] The UK has committed a budget of GB£3.87 billion to the International Climate Fund (ICF) between April 2011 and March 2015, of which around half is allocated to adaptation and the remainder to low-carbon development and forestry. Together, the multilateral development banks provided US$3.7 billion in adaptation finance in 2011 alone.[4]

Yet this represents only a small fraction of total development assistance. For example, official development assistance (ODA) reached around US$130 billion per year from OECD DAC (Development Assistance Committee) countries over the period 2010–12.[5] It is crucial to ensure that these core programmes are resilient to future climate change. Klein (2001) describes three ways in which climate change could affect development projects:

- *The direct risk to the expected long-term outcome of projects*. Certain projects are particularly sensitive to climate change, like those involving water supplies, food, natural resources management, human health, and disaster resilience. In addition, capital investments, such as roads, bridges, major irrigation systems, and dams, last for many decades and so may have to operate under a set of climatic conditions for which they were not designed.
- *The indirect risk to the expected outcome of the projects*. Climate change alters the natural, social, economic, and political environment in which projects operate. So an impact in one part of the world may have significant implications for another though global supply chains.
- *The effects of the project and its outcomes on the vulnerability of communities or ecosystems* to climate change. Many projects can have co-benefits that help reduce vulnerability to climate change but there can also be unintended negative consequences; projects can influence the vulnerability of communities (potentially irreversibly) well beyond the formal end of the project. This can potentially lead to a situation where short-term interventions result in *"long-term maladaptation, increasing vulnerability to climate shocks"* (Brooks, Grist, and Brown 2009, p. 741).

"Climate-proofing" involves designing and implementing projects in such a way that they achieve their desired objectives (outcomes) irrespective of current variability and future climate change and avoid any negative impacts on the long-term vulnerability of people or economies (Klein et al. 2007; OECD 2009). This is a material issue for development organisations (Klein et al. 2007). For example, a review in 2006 concluded that 40% of World Bank projects were at significant risk from climate (World Bank 2006). An OECD analysis assessed all official aid flows from 1998 to 2000 to six developing countries and found that US$ half a billion per year in flows to Bangladesh and Egypt, and about US$200 million to Nepal and Tanzania, were at risk from climate change (van Aalst and Agrawala 2004). In 2006 OECD member states made a commitment to integrate adaptation into development cooperation (OECD 2006). This is now a public commitment made by many development organisations. For example, in DFID's 2011–2015 Business Plan there is a commitment to make programmes more "*climate-smart*" (DFID 2011a).

However, a critical question for aid agencies is where *additional* action is required *today* to manage expected climate-related risks to projects in the *future*. In this paper, we refer to this as "future-proofing".[6] Investing in future-proofing today is not necessarily the best course of action in all cases as it can entail greater costs or trade-offs to secure benefits that may not be realised for a decade or more. For example, the World Bank (2006) estimated that accounting for future climate in high-risk projects today could potentially increase project costs by between 5% and 15%.[7] Dercon (2012) and Béné et al. (2012) highlight that adaptation can in some cases entail a productivity trade-off. These costs and trade-offs of future-proofing must be weighed against the urgent need to allocate resources where they can have the greatest impact on poverty reduction today.

This article sets out to identify where future-proofing might be justified as an immediate action. Our central hypothesis is that although many projects could be deemed to be at risk from climate change (e.g., that 40% of World Bank projects), in only a small number of cases is *additional* action justified *today* to manage expected *future* climate change. The article sets out to test this hypothesis through the application of a framework to identify projects where future-proofing might be justified today.

In the second part of our analysis, we ask where there are signs that action may be justified today how well this is identified by development organisations? Finally, we provide best practice examples of different types of future-proofing from recent development projects. Throughout, we apply frameworks that are simple enough to be used routinely by aid agencies to identify projects requiring additional analysis.

In the following section, we introduce the methodology for the study, including the proposed framework for identifying where future-proofing might be justified. We then give the findings from the screening exercise for three countries and two development organisations, before considering *how projects should be designed differently* to account for long-term risks today and drawing examples of best practice from recent programmes. Finally, we discuss the findings, with a focus on the barriers to future-proofing in practice.

Methodology

Conceptual framework for future-proofing

The framework aims to identify where *additional* future-proofing action is likely to be justified today despite the potential costs and trade-offs. To do this fully would require detailed cost-benefit analyses. We introduce a simple qualitative approach to screen projects based on the frameworks outlined in Fankhauser, Smith, and Tol (1999).

Fankhauser, Smith, and Tol (1999) lay out an economic framework for appraising the *optimal timing of adaptation*; where there are benefits to adapting now to future climate and where this can be left until later given the additional costs that may be involved.[8] They conclude that early action is likely to be justified in the case of "*quasi-irreversible investments with a long lifetime (e.g. infrastructure investments, development of coastal zones)*", where they suggest that "*precautionary adjustments may be called for to increase the robustness of structures, or to increase the rate of depreciation to allow for earlier replacement*" (p. 67). Another important area identified by Fankhauser et al. (2013) is where there are *long-lead times* for action. This includes for example, research and development, building institutional capacity, and migration out of hazard-prone areas.

In the context of development projects these studies suggest three areas where additional action today is justified to adapt to future risks:

(1) *Long-lived, investments with large sunk costs*, such as hydropower stations, roads, dams, and other infrastructure. A failure to account for climate change upfront in such long-lived investments could mean that they underperform (e.g., in the case of water supply systems and hydropower) or become exposed to increasing damage. This could mean that investments need to be retrofitted or replaced prematurely, imposing greater costs. Figure 1 illustrates the lifetime of different investments; new transport and energy infrastructure can last for 40 years or more, large dams for at least 60 years, and patterns of urban development (the layout of suburbs, roads, and other infrastructure in a city), for more than 100 years. The climate is likely to be very different on these timescales. Capital investments are particularly prone to maladaptation because they tend to be difficult to change over time.

(2) *Long-term planning and policy-making*, such as growth strategies, sector development plans, a poverty reduction strategy, coastal development plans, drought contingency plans, and urban zoning can have far-reaching and complex consequences that influence vulnerability for decades. In some cases, they will have positive co-benefits for long-term resilience, for example, through strengthening governance, building capacity, and increasing access to credit. But in a few cases there is a risk of maladaptation when people are inadvertently committed to greater and difficult-to-reverse decisions that may increase the risk from climate change. This includes:
  • Social protection systems can increase resilience to climate shocks but will need to be adjusted over time to cope with the changing profile of vulnerability and climate risks.

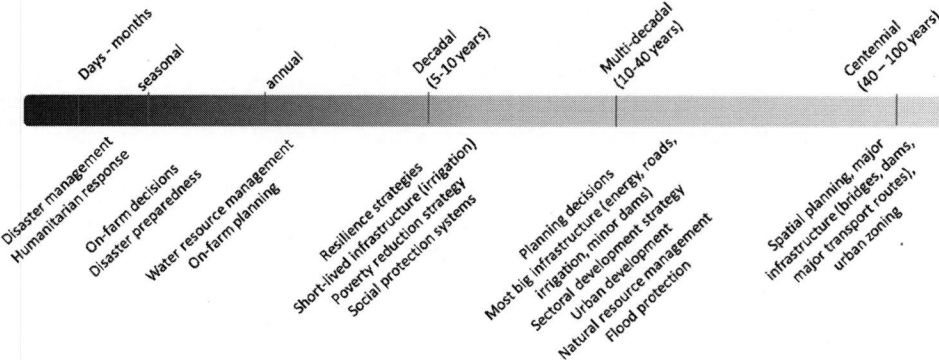

Figure 1.   The timescales of different types of climate-sensitive decisions.
Note: Based on Stafford-Smith et al. 2011.

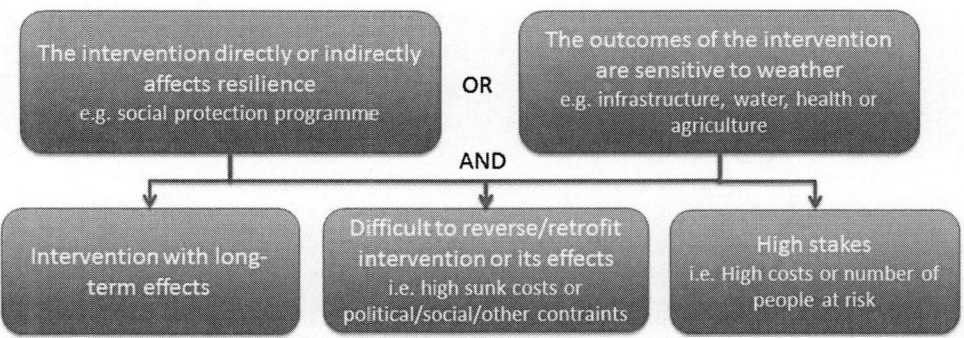

Figure 2.   Simple framework illustrating the conditions under which long-term climate change is likely to be an important factor in the design of a programme.

> If this adaptability is not built in from the start, systems can be difficult to adjust over time, due to political, social, or legislative barriers, making them less effective.
> - A programme that promoted water-intensive agriculture may change behaviour semi-irreversibly and be detrimental if the climate became drier (IEG 2012).
> - A rural roads programme that built intersections on floodplains could lead to urban development and put these communities at risk in the long term (IEG 2012).
> - A project that built schools on a floodplain could, at best, limit access to education for local children, or at worst, put them in danger (Save the Children 2008).
> - Even short-lived projects, like climate-smart agriculture or rural development programmes, can cumulatively add-up to major changes in long-term resilience in unexpected ways.

(3) *Interventions with long lead-times*: in cases where measures will take many years to implement, it may need to start now. For example:
   - Removing barriers to adaptation and building adaptive capacity can take time, as it can involve major changes in institutional, governance, and legislative structures (e.g., land and water rights), decision processes, and cultural norms and behaviour.
   - Research and development, for example, to develop and pilot new agricultural technologies can also take many years.
   - Changing livelihoods and migration, for example, enabling rural communities in unsustainable areas to move and seek new economic opportunities can take time.

This leads to a set of criteria that can help in identifying projects where it may be beneficial to future-proof now. In general, where the project or its outcomes are long-lived (i.e., long-term), difficult-to-adjust, and have a high cost or impact (i.e., high stakes) then climate change is likely to be a central factor in design today (Ranger 2013). This framework is illustrated by the lower three blocks in Figure 2. Conversely, where the project or its outcomes are short-lived, low-cost, or adjustable over time,[9] then accounting for long-term climate change is less likely to be a central factor in design (Hallegatte 2009). A similar criterion for the urgency of adaptation was used in the UK's National Climate Change Risk Assessment (Defra 2012).

Screening methodology

To test our central hypothesis we apply our future-proofing framework to the project portfolios of two development organisations, the World Bank and the United Kingdom's Department for International Development (DFID), for three countries over the period January 2007 to September

2013.[10] All information is drawn from publicly available sources.[11] Altogether we evaluate almost 250 projects with a total value in excess of US$4.5 billion. However, this is a small slice of total development assistance provided internationally to the countries in question. The limited scope of the sampling means that it is not possible to generalise our conclusions, but it does provide an initial view to guide further analysis and test the hypothesis posed by this paper.

The three countries selected are situated in East Africa (country A), South Asia (B), and the Caribbean (C).[12] The East African and South Asian countries are low income and the Caribbean country is middle-income. The choice of country was guided by those considered to be vulnerable to climate change and where there is clear exposure to extreme weather events today (with several major events in the past decade). We selected countries with mid-sized portfolios (e.g., around 50 DFID projects over 2007–13) to give a good sample size. We excluded countries seen as leaders on climate change (e.g. Ethiopia and Bangladesh) to avoid biasing our sample. The three countries selected have very different geographies but each country is exposed to flooding as well as other forms of natural hazards.

Before applying the future-proofing framework, we conducted a risk screening of all the projects in the country portfolios of the two agencies to identify those with a potential sensitivity to climate. The aim was to produce a set of "climate-sensitive" projects similar to the 40% identified by World Bank (2006). The risk screening was undertaken through a simple tool that combines elements from Burton and van Aalst (2004), EuropeAid (2009), and DFID (2012a).[13] This measures two elements of sensitivity: (1) the direct or indirect effect of the intervention on the resilience of people or systems, and (2) the sensitivity of the outcomes of the intervention to weather. This is shown by the top two blocks in Figure 2. We note that this screening can only tell us where there is a *potential* risk (Hammill and Tanner 2011) as the method does not take any account of the quality of project risk management. The future-proofing framework was then applied to projects scored as having a medium or high sensitivity to climate. Unless otherwise stated, all results are given in terms of numbers of projects rather than their value.

We emphasise that this analysis is preliminary and should be seen in light of the following qualifications. First, the findings cannot be extrapolated to all developing countries and the projects may not be representative for either organisation. Second, we do not consider additional activities within the country, including changes to development plans and the role of non-traditional donors or a broader spectrum of development activity. Third, this evaluation is based on publicly available documents that provide a limited picture of the projects and the organisations in question. For these reasons, this evaluation should be considered a snapshot that should be deepened with further analyses.

Evaluating current action

The final part of the analysis evaluates if and how climate is considered in the project design of medium to high potential risk projects through a review of publicly available documentation. Projects are rated as reaching one of five levels:

Level 1. No mention of climate or climate change at all
Level 2. Climate risks mentioned in the documentation (but not climate change)
Level 3. Climate change mentioned in the documentation
Level 4. Risks from climate change to the project and/or opportunities discussed
Level 5. Signs that future-proofing is integrated in the project design.[14]

Three constraints on this analysis are noted. First, public records are incomplete for some projects, particularly in the DFID portfolios in the earlier years; these projects are not included in the analysis. This means that the sample sizes for this part of the analysis are smaller. Second, even if climate risks are mentioned in the documentation, this analysis cannot evaluate whether this was implemented in practice and the quality of the implementation.[15] The reverse is also true; if climate risks are not mentioned in documentation, the organisations commissioned to implement the projects may still take action to future-proof projects. This analysis can only tell us, therefore, if risks are being identified and managed by development organisations in the early stages of project design. Third, we find that the description of future-proofing measures in the project documentation is often highly generalised (e.g., they often refer to mainstreaming but give no details on what this will entail specifically) and so it is difficult to judge the true level of integration – in this analysis, we tend to give the benefit of the doubt, so there may be a positive bias.

Screening results
Sensitivity of portfolios to climate risks
In agreement with previous studies, we estimate that around 30% of the projects (by value) have a medium or high potential risk, although this varies by country and portfolio from about 20% to 80% (Figure 3). This is driven by the risk profile of the country and the types of projects within the portfolio rather than being a reflection of the quality of risk management. About 40% of projects (by value) were rated as negligible risk.

For country C (Caribbean), about 80% of projects (by value) are rated as high potential risk. This could be because natural hazards and climate change are some of the greatest risks facing this country and therefore development projects tend to focus on climate-sensitive sectors. It is also a much smaller portfolio and the findings are skewed by a few big projects.

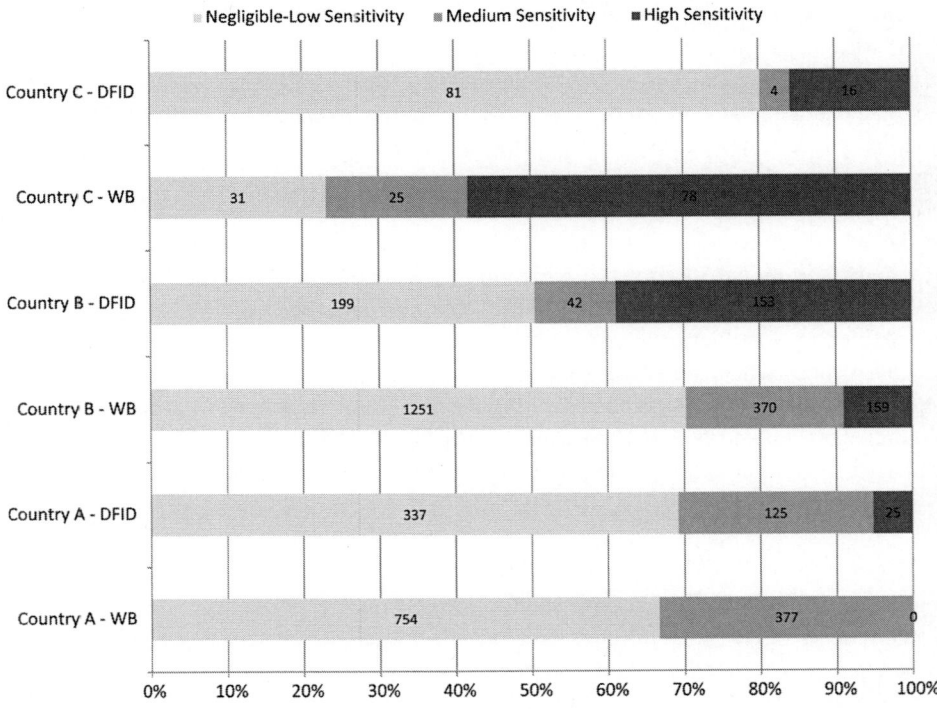

Figure 3.   Screening of the potential sensitivity of the project outcomes to climate change.
Note: Values indicate numbers of projects.

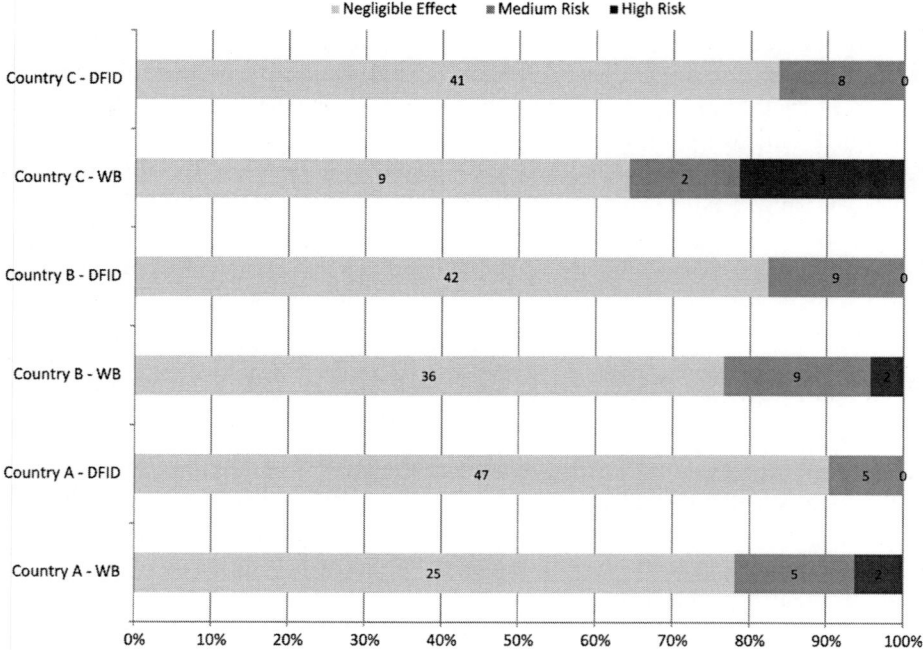

Figure 4.   Screening of the potential for a negative impact from the project on local resilience.
Note: Values indicate numbers of projects.

The highest potential risk projects across all countries are mainly major infrastructure projects (hydropower, bridge maintenance, water supply infrastructure, and post-disaster reconstruction projects) and weather risk insurance. The DFID portfolio also includes a number of climate change adaptation projects that by their nature address sectors that are sensitive to climate. Medium-rated projects tended to be major agricultural programmes, social safety net programmes, regional electricity transmission line projects, water resource management, and rural roads development projects.

Figure 4 shows that between 10% and 35% of projects could potentially have a negative impact on local resilience if climate change is not accounted for properly; this represents 30% of the portfolio across the three countries and two organisations by value and over US$1 billion. This includes many of those medium-to-high risk projects in Figure 3 (e.g., the failure of a hydropower station due to climate change would threaten local resilience), but also additional projects, such as support to extractive industries (which can both enhance and reduce resilience at different scales)[16] and urban governance (which affects urban resilience). This result should be balanced against our finding that over 80% of projects have a *strong* potential to increase long-term resilience; through, for example, building institutional capacity, increasing wealth and improving health, education and financial services (addressing the *"adaptation deficit"* [Fankhauser and McDermott 2013).

Where might future-proofing be justified today?

Figure 5 suggests that a significant proportion of projects rated as having medium or high potential risk could require urgent action to future-proof activities against climate change. This varies significantly between countries, ranging from about 5% to 55% of projects (or 2% and 30% of all projects).

Figure 6 summarises the types of projects that tend to emerge as *"high urgency"* in the analysis. The most common are public buildings (schools, hospitals) and transport infrastructure, followed by urban planning, infrastructure, and post-disaster reconstruction.

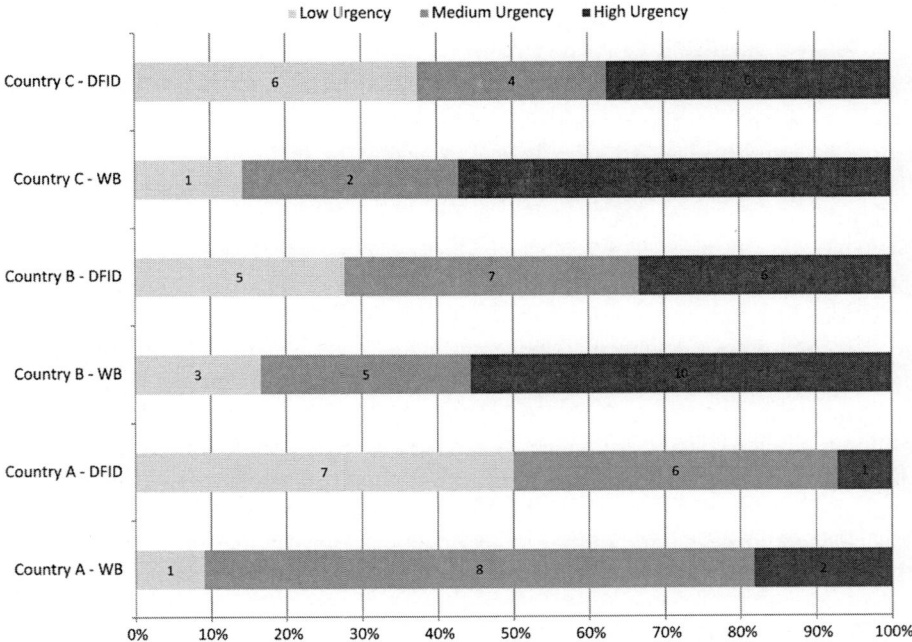

Figure 5.   Urgency of considering long-term climate risks in development intervention for those projects at medium to high risk.
Note: Values indicate the numbers of projects.

In the main these findings support our central hypothesis that in only a small number of cases would *additional* action be justified *today* to manage expected *future* climate change. However, this hypothesis does not hold for all country portfolios. In particular, country portfolios with more major infrastructure projects (as in the World Bank portfolios) tend to require more urgent future-proofing.

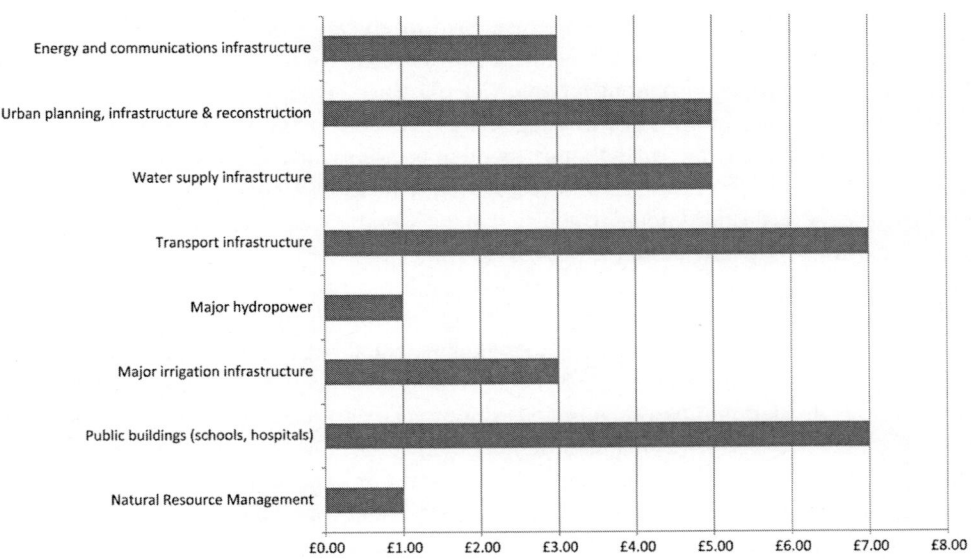

Figure 6.   Types of projects rated as requiring urgent action.

Are climate risks recognised in project documentation?

Across the DFID portfolio we find that over 90% of medium to high potential risk projects at least mention climate change (i.e., rated Level 2 using the method outlined in Section II.c) and around 50% include some form of future-proofing to climate change (Level 5) (Figure 7). However, many of these projects are targeted climate change projects. This includes, for example, projects to establish national climate change funds, support the development of climate change strategies, and "climate-proof" infrastructure.

If all International Climate Fund (ICF) projects (i.e., those DFID projects specifically targeted at climate change adaptation and mitigation) are removed from the sample, then the proportion of projects that include some explicit future-proofing falls to just over 30%, although recognition of climate change in project documentation is still high at 88%. Examples of future-proofing here include: incorporating a monitoring strategy into a water supply project to detect signs of climate change; mainstreaming climate change into national planning; capability building and establishing a climate change committee as part of a social protection programme. The extent of future-proofing is variable between countries. For example, for Country C, no non-ICF projects at medium-to-high potential risk incorporate future-proofing, while for Country A half of the non-ICF projects incorporate future-proofing. This finding requires further investigation before one could draw conclusions. For example, it may be a strategy of Country C to put all climate-sensitive projects into the ICF.

Importantly, many projects that are likely to require future-proofing today, like water and sanitation projects, rural road projects, and disaster resilience, are not ICF projects. If we focus our sample on only these projects (i.e., non-ICF, medium-high sensitivity and likely to require future-proofing), then we find that around one third include future-proofing explicitly. However, the sample size with available documentation is small – only six projects.

Across the World Bank portfolio, we find that around 20% of medium-high potential risk projects mention climate change. Across the 93 projects reviewed we identified six where climate change was explicitly considered in the project design. This included, for example, climate-proofing social protection systems, post-disaster integration of climate change adaptation into reconstruction, and incorporating knowledge generation and capacity building into projects. But performance varies significantly by country; for country C, all projects at least mention climate risks (Level 2) and almost half integrate resilience to climate change into the project design (Level 5). This might be a reflection of the political economy of the county in question that would influence the scope of options proposed for IDA assistance.

Of those World Bank projects that we suggest are likely to require future-proofing today (16 projects in total), we find only one where future-proofing measures are explicitly incorporated into the design documentation. However, the sample size is too small to draw conclusions. The recent World Bank Independent Evaluation Group's evaluation of the World Bank portfolio (IEG 2012) conducted a more extensive review and concluded that there is a general lack of forward-looking, pro-active development projects, which anticipate future risks and act to reduce them ahead of time.

There are clear signs of improvement in integration of climate change in both portfolios. Figure 8 shows the proportion of all medium to high potential risk projects, where documentation is available,[17] that at least reference climate change in their documentation (i.e., reach at least Level 3 above) divided into two time periods (by project start date): 2007 to 2010 and 2011 to 2013. We find that all of the medium-high potential risk DFID projects at least reference climate change in the latter period and the proportion of World Bank projects mentioning climate change improves from 14% to 22%. This estimate is consistent with the IEG evaluation of the World Bank's experience on mainstreaming adaptation (IEG 2012), which found the proportion of all World Bank projects mentioning climate change adaptation increased from less than

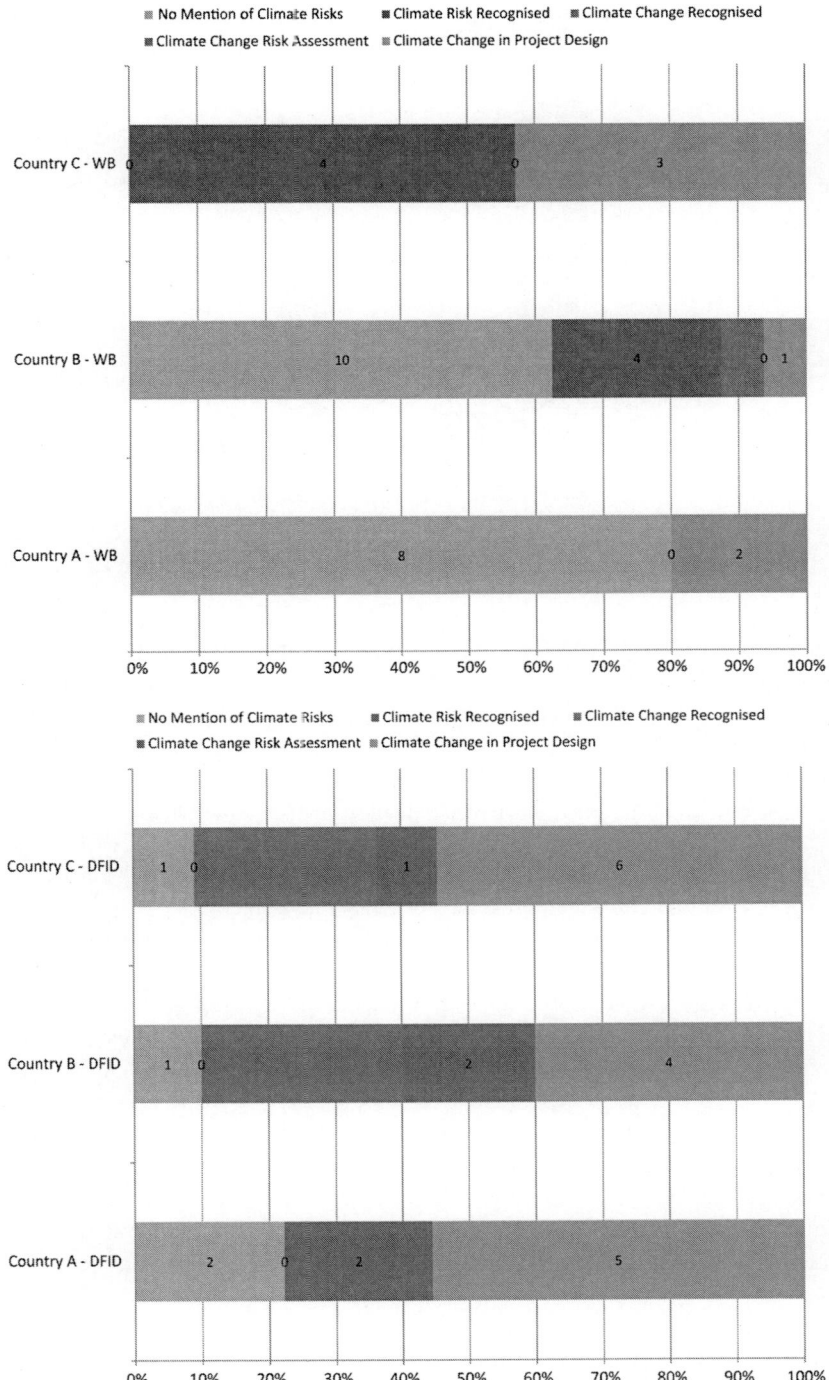

Figure 7. Level of integration of climate risks and climate change in the project documentation of medium and high risk projects: (top) World Bank portfolio and (bottom) DFID portfolio.

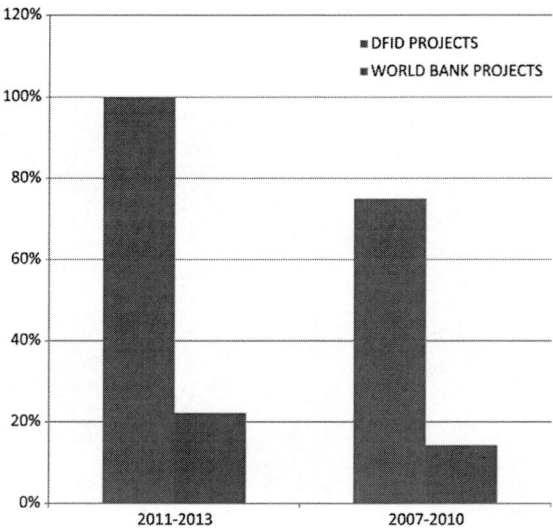

Figure 8.   Proportion of projects rated as medium to high potential risk that at least reference climate change in the publicly-available documentation.

1% in 2000 to around 12% by 2011. This is a clear improvement from situation pre-2007 based on Klein et al. (2007), which reviewed the portfolios of six development agencies and concluded that climate risks were not well assessed and rarely mentioned in project documentation.

The higher level of integration within the DFID portfolio may partly reflect the introduction of the Climate and Environment Appraisal (CEA) process in late 2010. This made it mandatory for all DFID project business cases to include an assessment of the potential risks and opportunities from climate change to the project. DFID's Strategic Programme Review (2010/11), International Climate Fund and "Future Fit" initiative (2012/13), may also have driven greater awareness and incentives for integration. Over the period of this study, assessment of climate change risks was not mandatory within World Bank projects. In addition, the World Bank is a far larger organisation, so integration at the project level will inevitably take longer to achieve. Given that climate change is a special theme for IDA 17, we expect the level of integration to increase over the coming years.[18]

For all these findings, we reiterate our earlier qualification that just because climate change is integrated into the project documentation, this does not necessarily mean that the project is successful in building long-term resilience. This is an assessment of *process* rather than *outcomes*. In addition, we have found that the discussion of future-proofing in project documentation tends to be highly general and so it is difficult to draw robust conclusions on the level of integration (e.g., often discussing mainstreaming without giving details). More detailed study is required to assess the true extent of integration of future-proofing into the project design and, more importantly, the project outcome. Such a study would require on-the-ground evaluation of projects.

Identifying best practice in future-proofing

This final part of the analysis aims to identify examples of best practice in future-proofing across the portfolios of the two development organisations. To do this, we propose a framework for what future-proofing would look like in practice for development projects, based on those proposed by Klein et al. (2007), McGray et al. (2007), and OECD (2009).

Each of these three studies suggests that future-proofing does not necessarily mean that projects will look radically different or that they will be more expensive. Klein et al. (2007) stress that

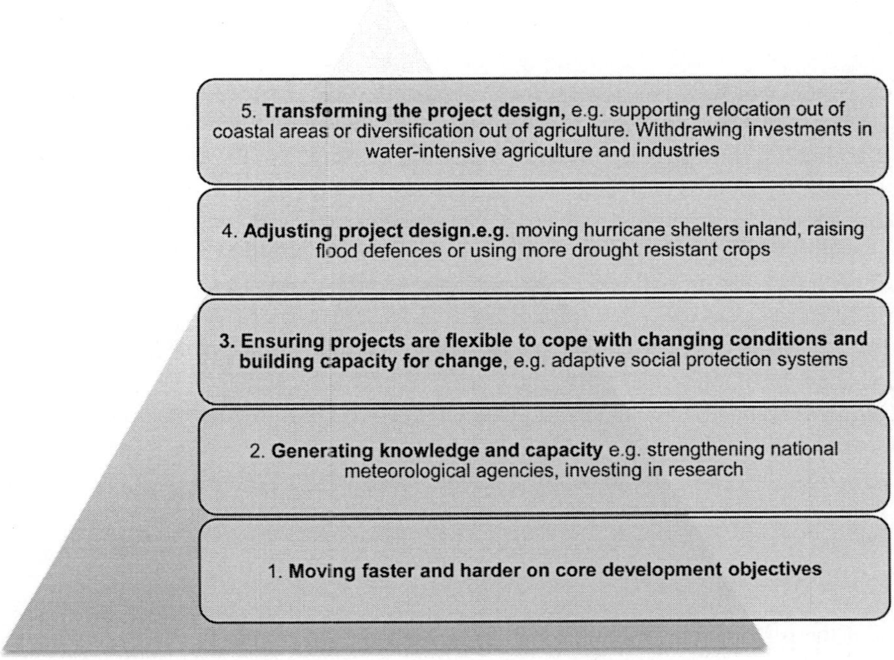

Figure 9.   An illustrative scale of how the inclusion of climate change within interventions today may alter those interventions, relative to the counterfactual.

future-proofing is not necessarily about technological measures, like building bigger pipes in drainage programmes or drought-resistant crops, but can also include training and capacity building and institutional support. McGray et al. (2007) suggest that adaptation can entail a continuum of measures, from "*low-regret*" measures like reducing vulnerability and capacity building, through to making more significant adjustments to plan to explicitly address future risks, such as relocation and radical technological change (OECD 2009). This is consistent with our findings from the review of the portfolios from three countries.

We suggest that a similar continuum applies to future-proofing development projects. For some projects, low-regret measures may be sufficient, but a few may need to move further up to continuum toward major adjustments. Figure 9 illustrates this continuum; the pyramid structure reflects our hypothesis that there will be fewer projects requiring more radical adaptations. In this section, we describe our proposed continuum and give best-practice examples from the wider portfolios of the World Bank and DFID.

In line with McGray et al. (2007), we suggest that the first and second levels of the continuum of integrating *long-term* climate change are: increasing efforts to reduce vulnerability through good development and reducing other stresses; and building *general* capacity, including investing in knowledge and skills. Fankhauser et al. (1999) showed that long-term climate risks increases the rationale for moving *faster and harder* on core development and disaster resilience, and for investing in information. From our review of the portfolios of three countries, we find that most projects (more than 80%) are likely to have positive impacts on resilience to climate change.

An example is the "Adaptation for Smallholder Agriculture Programme" (ASAP) (DFID 2012b), launched in 2012 by the International Fund for Agricultural Development (IFAD). ASAP works in 40 countries to invest in practices and knowledge to build the capacity of smallholder farmers to manage current risks and adapt to climate change.[19] It aims to safeguard the food systems on which poor smallholder farmers depend and to demonstrate how to scale up practices and technologies that build resilience and increase prosperity.

Another example is the Ethiopian Productive Safety Net Programme (PSNP) Climate Smart Initiative (CSI) (World Bank 2013b). The PSNP provides food and cash to food insecure Ethiopians; in a typical year 14 million Ethiopians do not have enough food to eat and climate variability is an important underlying driver.[20] To address chronic food security over the long term, Ethiopia needs to build a consideration of climate change into its current food security programming. The PSNP CSI will enhance adaptive capacity through improving information flows, such as weather information and early warning systems, and supporting local decision-making in the PSNP. It will also help to build capacity to adapt for the long term by enabling communities to diversify livelihoods and promote technological innovation.

The third level is where projects begin to look very different. The focus here is not on making radical changes now, but on designing the project so that it can anticipate, learn, and evolve over time to cope with changing climate risks, while still achieving its objectives (Walker et al. 2013). A commonly cited example is the Thames Estuary 2100 Project, where instead of building a costly new flood barrier for London today, a plan was developed that allowed the flood management system to be adjusted incrementally over time as more is learnt about future risks. The same principles can be transferred to development interventions (Ranger 2013); an example is an *adaptive* social protection system, which is designed from the start to expand or contract in response to changing needs (IDS 2012). A number of the projects reviewed from the three sets of portfolios included features that aimed to build adaptive capacity. For example, one project recommended a monitoring system as part of a water supply programme to detect early signs of climate change.

These first three levels are low-regret; they are low cost and likely to have benefits today and in the future (OECD 2009). The next two levels entail measures that are designed explicitly to account for future risks now. Designing these types of projects will likely require more technical expertise and information, including detailed analyses of risks under different climate and socio-economic scenarios and quantitative assessments of options. This is likely to apply to fixed, long-lived capital investments that are difficult to adjust.

The fourth level is akin to Klein et al.'s (2007) building bigger pipes in drainage programmes or drought-resistant crops. It can mean changes in engineering (such as higher sea walls, better drainage systems, resilient school buildings), new practices (climate-smart agriculture), new policies (water permits, resilient urban zoning, and stronger building codes), and different technologies (rainwater harvesting, insurance) (OECD 2009). Two examples from across the whole portfolio of the two organisations are:

- The Haiti Strategic Programme for Climate Resilience (CIF 2013), which conducts a detailed assessment of climate change risks and proposes a strategy to engineer its infrastructure, agricultural systems, and coastal cities to cope with future risk, as well investing in measures to build adaptive capacity through climate monitoring, training, and institutional strengthening.
- The US$412 million Trung Son Hydropower project in Vietnam that adopted high safety standards, zoning and warning systems, and built a secondary "*fuse dam*" to reduce the risks associated with failure due to changing river flows (IEG 2012). Several projects within the three sets of portfolios reviewed referred to climate-proofing programmes and infrastructure, and mainstreaming climate change into policies and plans.

Finally, the fifth level is the potential transformation of a project where a major change is required. For example, Stafford-Smith et al. (2011) describe how in the highest risk areas, communities may need to radically change where and how they live to cope with climate change. One example is a small-island state, where populations may need to relocate to escape rising sea levels. We suggest that in some cases, development projects may need to similarly transform, or drive transformational change in societies.

We found few examples of where this has been done in practice in the literature and none from the three sets of portfolios reviewed. An example is a recent World Bank project in India (reported by IEG 2012). The Indian portion of the Sundarbans, a great mangrove-dominated delta facing the Bay of Bengal, is home to more than four million poor people. Many live at or below sea level and are at constant risk from floods and cyclone. An analysis found that many well-intentioned and apparently adaptive activities, like strengthening embankments, were maladaptive, boosting long-term vulnerability. While there are urgent poverty challenges it concluded business-as-usual development is not sustainable in the long-term. Instead, the project proposes that the Sundarbans embark on a multi-generational plan to re-engineer estuary management (e.g., moving back defences and allowing mangroves to recover) and enable welfare-improving voluntary out-migration from the most threatened areas, including through education. In the short term, this would be complemented by investments in early warning systems; cyclone shelters; and health, water, and sanitation services (IEG 2012). These types of programmes can be extensive and entail difficult and complex trade-offs.

Discussion

This study has suggested a relatively low integration of future-proofing strategies into the recent projects of two development organisations for three countries since 2007. While it is not possible to extrapolate these findings to other countries and organisations in this study, other studies have reached similar conclusions (e.g., IEG 2012).

Given this, it is important to consider what the barriers are to integrating climate change into development programmes. Practical experience within development agencies, such as the two described in this paper, suggests that there are many. This is also reflected in the existing academic literature. For example, Hammill and Tanner (2011) report that the difficulties in identifying and assessing climate and complex vulnerability information is often cited as one of the biggest challenges here. There is also a lack of information about the costs and benefits of different adaptation measures and until recently, little robust or systematic monitoring and evaluation of adaptation.

However, other literature suggest more fundamental political, structural, financial, and behavioural barriers. Historically, planning and policy-making is often slow to react to changes in the external environment and institutions have limited capacity to learn from and foresee change. Jones et al. (2013) concludes that the majority of climate/disaster resilience and humanitarian projects tend to be either reactive (managing events as they happen) or deliberative (learning from the recent past and adapting to it). The challenges to future-proofing may include[21]:

- There may be a general lack of awareness of the issues.
- Lack of experience in future-proofing amongst development organisations – this is a new agenda and learning is required so it will take time to implement fully.
- Future-proofing will require more time, resources, information, and technical capacities, in an environment where there are already constraints.
- The administrative separation of finance for specific and additional adaptation from normal development programme finance may also create a barrier in this respect.
- The short duration of many development projects (only around three to five years).
- The incentives on development professionals for projects that deliver rapid impacts and high returns on investment will tend to reduce investment in adaptation, which is often percieved as having more uncertain, long-term, difficult-to-quantify, and sometimes intangible benefits.
- A lack of long-term monitoring and evaluation systems for adaptation.

Another often cited challenge in integrating future climate into decisions today is the uncertainty over long-term climate projections (IPCC 2013). Most risk assessments are based on historic data; but planning for climate change requires a more forward-looking approach. The risk of getting it wrong increases as one moves up the continuum of adaptation. For the first three types of adjustments (Figure 9), uncertainty is unlikely to be important as the measures are *"low-regret"* (OECD 2009). Where major changes to plans are made to account for long-term climate change, the potential for maladaptation is greater. Uncertainty means that we cannot optimise the design of a project to a particular future climate. However, in theory this should not be a barrier to integrating resilience to climate change into development projects (Ranger 2013). A desirable approach is one that is *"robust, meaning that it performs well under a wide variety of futures, and adaptive, meaning that it can be adapted to changing (unforeseen) future conditions"* (Walker *et al.* 2013, p. 956). There are tools available to help do this in practice, for example so-called futures techniques like scenario planning and methods for decision making under uncertainty (Ranger 2013).[22]

A further issue is the potential trade-offs between long-term resilience and short-term poverty alleviation (Dercon 2012). For example, more drought-resistant crops tend to have lower productivity. Another example is the Sundarbans case given above. In that case a balance was struck by both planning for the long term by relocating some people and investing in building resilience and reducing poverty in the near term. These so-called wicked problems in adaptation require a high level of understanding of the complex societal processes that generate poverty and vulnerability (Klein *et al.* 2007; Jones *et al.* 2013).

What more can be done to enable and promote future-proofing of development projects? Appropriate evidence and tools are a foundation to this; for example, there is a need for further economic analysis to identify where and how long-term risks should be built into projects today and to provide learning examples. Training and skills will also be important in applying evidence and tools in practice. But the most important advance must be to build an institution that creates the right incentives to integrate climate change; this requires leadership to institute a cultural change that places a greater value on the long-term sustainability of development investments and progress.

Both organisations are working to deliver both specific climate change interventions and also mainstream climate change across other areas through portfolio screening or safeguard systems. For example, the DFID Future Fit programme initiative aims to integrate long-term resilience to climate change and resource scarcity across all DFID programmes. Coupling climate change resilience to the disaster resilience agenda (DFID 2011b), provides an opportunity to do this more efficiently.[23] For the World Bank, climate change is a special theme for IDA 17 and the Bank has committed to integrate climate risk into all new developments (World Bank 2013c). It is also worth noting the Management Response to the recent IEG evaluation of World Bank experience with adaptation.[24] We recognise that changing the operations, practices, and behaviours of large institutions takes considerable effort, time, and leadership. This evaluation should therefore be considered an initial snapshot and should be repeated after IDA 17 and broadened with further analyses.

Conclusions

This article gives a framework for identifying where and how development projects should integrate long-term climate risks into their design today. We conclude that portfolio-level screening can be a useful tool, but more project-specific evaluations are needed to assess the extent to which long-term climate is being integrated into projects in practice and the barriers faced by development professionals in this area.

Acknowledgements

We thank Sarah Arnold for her assistance in the portfolio analysis and John Carstensen, Jane Clark, Stéphane Hallegatte, Richard Teuten, and Tim Wheeler for their comments. Any remaining errors are our own. Nicola Ranger's research was supported by the Global Green Growth Institute (GGGI), the ESRC Centre for Climate Change Economics and Policy and the Grantham Foundation for the Protection of the Environment. The views expressed here are those of the authors and do not necessarily reflect those of their institutions.

Notes

1. Data provided by Munich Re.
2. Adaptation can be defined as *"adjustments in human and natural systems, in response to actual or expected climate stimuli or their effects, that moderate harm or exploit beneficial opportunities"*: IPCC (2012).
3. UNFCCC Document FCCC/CP/2009/11/Add.1 (paragraph 8)
4. Joint MDB Report on Adaptation Finance 2011: A report by a group of Multilateral Development Banks (MDBs) comprising the African Development Bank (AfDB), the Asian Development Bank (ADB), the European Bank for Reconstruction and Development (EBRD), the European Investment Bank (EIB), the Inter-American Development Bank (IDB), the World Bank (WB) and the International Finance Corporation (IFC). Accessed September 20, 2013. http://www.eib.org/attachments/documents/joint_mdb_report_on_adaptation_finance_2011.pdf
5. Aid statistics from OECD DAC. Accessed September 20, 2013. http://www.oecd.org/dac/stats/data.htm.
6. "Future-proofing" is defined here an *additional* action to anticipate *future* risks and act now to reduce them ahead of time. We emphasise "additional" because in some cases the measures to cope with future risks are the same as those needed for current risks, e.g., building institutional capacity. Future-proofing is one part of "climate-proofing", which includes taking action to manage existing climate risks. The distinction is important as managing existing climate risks often bring immediate benefits, whereas for "future-proofing" the full benefits will not be realised until later.
7. Such estimates should be interpreted with caution as the costs will vary significantly between projects.
8. This includes real costs, such as building a hydropower station so it that operates under a wider range of river flows, opportunity costs, and benefits foregone (e.g., from building on flood plains).
9. Short-lived, flexible, and lower-cost projects can provide major opportunities to build resilience to climate (Klein et al. 2007), but this is not a focus of this paper.
10. The start date, 2007, is chosen as we suggest that this roughly the point in time when the linkages between climate change and poverty alleviation became much clearer and more mainstream in thinking, following the release of the Fourth Assessment Report of the Intergovernmental Panel on Climate Change in 2007, the Stern Review on the Economics of Climate Change in late 2006, and the Gleneagles G8 dialogue (with climate change and poverty as its two priority issues) in 2005.
11. Accessed September 20, 2013. DFID Development Tracker. http://devtracker.dfid.gov.uk/ and World Bank Projects Database. http://www.worldbank.org/projects. Accessed September 20, 2013. For DFID projects, the business case is the main source of information used and if not available, the most recent Annual Review or project completion report. For the World Bank projects, the Project Information Document is the main source of information use, though Project Appraisals, Environmental Assessments, and Implementation and Results Reports are also used.
12. The countries remain anonymous in this paper.
13. These rules are also consistent with guidance by USAID, GIZ, NORAD, and others (Hammill and Tanner 2011).
14. For example, the project might include measures to reduce climate risks or build capacity to cope with climate change, or climate risks may feature in the options appraisal.

15. For example, there it is difficult to establish whether climate change is merely name-checked or tick-boxed instead of being truly integrated into the project. This would require deeper investigation at project level.
16. Extractive industries could, in local terms, reduce resilience of the environment and local communities. However, at the national scale they could improve the ability of a country to manage external economic shocks (UNECA 2013).
17. For some DFID projects, particularly in the earlier period, no project documentation is publicly available.
18. IDA 17 is the next funding phase of the International Development Association (IDA), the part of the World Bank that focuses on helping the world's poorest countries.
19. These will include small-scale water harvesting and storage, flood protection, irrigation systems, agroforestry, and conservation agriculture, strengthening farmers access to markets and information (e.g., weather forecasts), and working with governments on policies to enable growth and climate resilience agriculture.
20. This is in return for labour for those who can provide it (around 80% of beneficiaries), and as a grant to those who are elderly and sick (around 20%) (DFID 2012b).
21. Challenges identified by the authors based on the literature review summarised by IPCC 2012.
22. See also, for example, the work of the Mediation programme (http://www.mediation-project.eu/platform/pbs/home.html) funded under the EU's 7th FP.
23. DFID conducted a number of activities to integrate disaster resilience into programmes, including pilots and risk assessments (DFID 2011b). A similar framework would be needed to deliver climate change resilience. Coupling the two processes together would therefore be more efficient.
24. World Bank, Washington, DC. "Management Response to the IEG 2012." Accessed September 20, 2013. https://ieg.worldbankgroup.org/Data/reports/chapters/cc3_mgmt_response.pdf

References

van Aalst, M., and S. Agrawala. 2004. "Analysis of Donor Supported Activities and National Plans." In *Bridge Over Troubled Waters: Linking Climate Change and Development*, edited by S. Agrawala, 61–83. Paris: OECD.

Béné, C., R. Godfrey Wood, A. Newsham, and M. Davies. 2012. "Resilience: New Utopia or New Tyranny?." IDS Working Paper 405. Brighton: IDS.

Brooks, N., N. Grist, and K. Brown. 2009. "Development Futures in the Context of Climate Change: Challenging the Present and Learning from the Past." *Development Policy Review* 27 (6): 741–765.

Burton, I., and M. van Aalst. 2004. *Look Before you Leap: A Risk Management Approach for Incorporating Climate Change Adaptation into World Bank Operations*. Washington, DC: The World Bank.

Climate Investment Funds. 2013. *Haiti Strategic Program for Climate Resilience*. Washington, DC: Climate Investment Funds.

Dercon, S. 2012. "Is Green Growth Good for the Poor?." World Bank Policy Research Paper. Washington, DC: The World Bank.

Defra. 2012. *UK Climate Change Risk Assessment: Government Report*. London: Department for Environment, Food and Rural Affairs. Accessed June 4, 2013. https://www.gov.uk/government/publications/uk-climate-change-risk-assessment-government-report

DFID. 2011a. *Business Plan 2011–2015 Department for International Development*. London: Department for International Development.

DFID. 2011b. *Defining Disaster Resilience: A DFID Approach Paper*. London: Department for International Development.

DFID. 2012a. "Climate Change & Environment Assessment for the Business Case." DFID Practice Paper. April 2012. London: Department for International Development.

DFID. 2012b. *Business Case for the Adaptation for Smallholder Agriculture Programme (ASAP)*. London: Department for International Development.

EuropeAid. 2009. *Guidelines on the Integration of Environment and Climate Change in Development Cooperation*. Brussels: EuropeAid.

Fankhauser, S. and T. McDermott. 2013. "Understanding the adaptation deficit: why are poor countries more vulnerable to climate events than rich countries?" Grantham Research Institute on Climate Change and the Environment, London UK, Working Paper No. 134.

Fankhauser, S., N. Ranger, J. Colmer, S. Surminski, and D. Stainforth. 2013. "An Independent National Adaptation Programme for Britain." Grantham Research Institute Policy Paper. London: Grantham Research Institute on Climate Change and the Environment.

Fankhauser, S., J. Smith, and R. Tol. 1999. "Weathering Climate Change: Some Simple Rules to Guide Adaptation Decisions." *Ecological Economics* 30 (1): 67–78.

Hallegatte, S. 2009. "Strategies to Adapt to an Uncertain Climate Change." *Global Environmental Change* 19: 240–247.

Hammill, A., and T. Tanner. 2011. "Harmonising Climate Risk Management: Adaptation Screening and Assessment Tools for Development Cooperation." OECD Environment Working Paper No. 36. Paris: OECD.

Independent Evaluation Group. 2012. *Climate Change Phase III - Adapting to Climate Change: Assessing World Bank Group Experience*. Washington, DC: The World Bank.

Institute of Development Studies. 2012. "Making Social Protection 'Climate-Smart'." IDS Policy Briefing No. 27. September 2012. Brighton: IDS.

IPCC. 2012. "Managing the Risks of Extreme Events and Disasters to Advance Climate Change Adaptation. A Special Report of Working Groups I and II of the Intergovernmental Panel on Climate Change." [Field, C.B., V. Barros, T.F. Stocker, D. Qin, D.J. Dokken, K.L. Ebi, M.D. Mastrandrea, K.J. Mach, G.-K. Plattner, S.K. Allen, M. Tignor, and P.M. Midgley (eds.)]. Cambridge: Cambridge University Press.

IPCC. 2013. "Climate Change 2013: The Physical Science Basis." Contribution of Working Group I to the Fourth Assessment Report of the Intergovernmental Panel on Climate Change. Accessed October 14, 2013. www.ipcc.ch/report/ar5/wg1/

Jones, L., E. Ludi, P. Beautement, C. Broenner, and C. Bachofen. 2013. "New Approaches to Promoting Flexible and Forward-Looking Decision Making: Insights from Complexity Science, Climate Change Adaptation and Serious Gaming." A report for the Africa Climate Change Resilience Alliance (ACCRA). London: Overseas Development Institute.

Klein, R. J. T. 2001. *Adaptation to Climate Change in German Official Development Assistance – An Inventory of Activities and Opportunities, with a Special Focus on Africa*. Eschborn: Deutsche Gesellschaft für Technische Zusammenarbeit.

Klein, R. J. T., S. E. H. Eriksen, L. Otto Næss, A. Hammill, T. M. Tanner, C. Robledo, and K. L. O'Brien. 2007. "Portfolio Screening to Support the Mainstreaming of Adaptation to Climate Change into Development Assistance." Tyndall Centre for Climate Change Research Working Paper 102. Norwich: Tyndall Centre for Climate Change.

McGray, H., R. Bradley, A. Hammill, E. L. Schipper, and J. Parry. 2007. *Weathering the Storm, Options for Framing Adaptation and Development*. Washington: World Resources Institute.

OECD. 2006. "Declaration on Integrating Climate Change Adaptation into Development Co-operation." Adopted by Development and Environment Ministers of OECD Member Countries. 4 April 2006. Paris: OECD.

OECD. 2009. *Integrating Climate Change Adaptation into Development Co-operation*. Paris: OECD.

Ranger, N. 2013. "Topic Guide. Adaptation: Decision Making Under Uncertainty." Evidence on Demand. doi: http://dx.doi.org/10.12774/eod_tg02.june2013.ranger

Save the Children. 2008. *In the Face of Disaster: Children and Climate Change*. London: Save the Children.

Stafford-Smith, M., L. Horrocks, A. Harvey, and C. Hamilton. 2011. "Rethinking Adaptation for a 4°C world." *Phil. Trans. R. Soc. A.* 369 (1934): 196–216. doi: 10.1098/rsta.2010.0277.

Stern, N. 2007. *The Economics of Climate Change: The Stern Review*. Cambridge: Cambridge University Press.

UNECA 2013. "Economic Report on Africa 2013: Making the Most of Africa's Commodities: Industrializing for Growth, Jobs and Economic Transformation." http://www.uneca.org/publications/economic-report-africa-2013

United Nations. 2013. "A New Global Partnership: Eradicate Poverty and Transform Economies Through Sustainable Development. The Report of the High-Level Panel of Eminent Persons on the Post-2015 Development Agenda." Accessed October 18, 2013. www.post2015hlp.org/the-report/

Walker, W. E., M. Haasnoot, and J. H. Kwakkel. 2013. "Adapt or Perish: A Review of Planning Approaches for Adaptation Under Deep Uncertainty." *Sustainability* 5 (3): 955–979.

World Bank. 2006. *Clean Energy and Development: Towards an Investment Framework. ESSD-VP/I-VP*. April 5, 2006. Washington, DC: The World Bank.

World Bank. 2010. *World Development Report 2010: Development and Climate Change*. Washington, DC: The International Bank for Reconstruction and Development/ The World Bank.

World Bank. 2013a. Turn Down the Heat: Climate Extremes, Regional Impacts, and the Case for Resilience. A Report for the World Bank by the Potsdam Institute for Climate Impact Research and Climate Analytics. Washington, DC: The International Bank for Reconstruction and Development/ The World Bank.

World Bank. 2013b. *Ethiopia's Productive Safety Net Program (PSNP) Integrating Disaster and Climate Risk Management: Case Study*. Washington, DC: The World Bank.

World Bank. 2013c. *"Special Themes for IDA17". IDA Resource Mobilization Department Concessional Finance and Global Partnerships June 2013*. Washington, DC: The World Bank.

Appendix: Climate Change Risk Screening Tool

The project screening conducted in this study aims to identify where there are *potential risks* to a project, based on a set of attributes and criteria about the programme. The attributes and criteria are taken from EuropeAid (2009), DFID (2012a) and Burton and van Aalst (2007) (which are all similar). It is a simple rules-based rating tool, similar to those currently used by DFID, USAID, ADB and GIZ (Hammill and Tanner, 2011). The tool used in this paper also screens where action may be needed today to enhance the resilience of the project to long-term climate change. These attributes, explained in Section II and Figure 2, are drawn from Defra (2012). The tool inevitably requires some professional judgement to apply. To reduce biases, we calibrate our ratings against the examples given by Burton and van Aalst (2007). The tool benefits from a basic level of knowledge about how climate change may affect the country(s) in which the project operates.

A. POTENTIAL DIRECT OR INDIRECT EFFECTS OF CLIMATE ON PROJECT OUTCOMES

Is the achievement of the programme's objectives directly and significantly dependent on the climate over the coming ten years or more?

Could the programmes objectives be indirectly affected by climate change? E.g. The achievement of objectives in a rural development programme may be highly dependent on the availability of increasingly scarce water for irrigation

High	*Possible major threat to long-term success of project* E.g. The projects is a large fraction composed climate sensitive sector (e.g. >50% agriculture, infrastructure or water) and entails climate sensitive activities (e.g. irrigation, water supply management, malaria).
Medium	*Outcome of project could be affected by climate, but unlikely to be a major threat to long-term success. E.g. Some elements of the project are subject to climate risks (e.g. >20% in climate sensitive sectors or with >10% climate sensitive activates) or exposed to indirect effects from climate change* (e.g. schools in risky areas). Also, projects that are climate-sensitive, but are likely to have benefits under any climate (e.g. social protection, capacity building in agriculture).
Low	*Climate change unlikely to be a threat to long-term success of programme.* Minor inclusion of climate sensitive sectors and activities OR potential indirect effects from climate change.
Negligible	Climate sensitive activities make a negligible fraction of the project

B. EFFECT OF THE PROJECT ON LOCAL OR REGIONAL VULNERABILITY

Are there any indications at this stage that the project could have positive impacts on vulnerability? (Y/N) e.g. Does the project enhance governance capacity?

Are there indications that the project may increase the vulnerability of the population to climate variability and/or the expected effects of climate change?

High	*Project could have a strong effect on the climate risks facing the country or region.* E.g. infrastructure could trigger development in dangerous areas, even if the infrastructure itself is not at risk.
Medium	*Project may have indirect effects on the vulnerability of communities.* E.g.an agricultural market reform project that removes subsidies from certain crops can lead farmers to switch to crops that could make them more vulnerable to climate variability and change.

C. URGENCY OF INCORPORATING CLIMATE CHANGE

(i) LIFETIME: How long will the outcomes of the project last: Long (>30 years), Medium (10–30 years), Short (<10 years).

(ii) FLEXIBILITY: How easy it is to change the activities to adjust over time? (Low, Medium, High). *E.g. Infrastructure is an inflexible investment, whereas early warning systems can be adjusted every year*

(iii) SCALE: How large is the project: Large (>50 m), Medium (>5 m), Small (<5 m).

Climate resilience in fragile and conflict-affected societies: concepts and approaches

Janani Vivekananda, Janpeter Schilling, and Dan Smith

To understand resilience to climate and environmental changes in fragile and conflict-affected societies is particularly important but equally challenging. In this paper, we first develop a conceptual framework to explore the climate-fragility-conflict and climate-resilience-peace nexus. Second, we discuss approaches to promote pathways from climatic changes to peace. We draw on the relevant literature and International Alert's experience in fragile and conflict-affected societies to stress the key role of resilience. To build resilience, climate, development, peacebuilding, and government actors would have to overcome bureaucratic and institutional barriers and cooperate across thematic and regional silos.

Comprendre la résilience face aux changements climatiques et environnementaux dans les sociétés fragiles et touchées par des conflits est tout particulièrement important mais tout aussi difficile. Dans cet article, nous élaborons en premier lieu un cadre conceptuel pour explorer le lien entre climat-fragilité-conflit et climat-résilience-paix. Nous traitons en second lieu des approches pour promouvoir les chemins menant des changements climatiques à la paix. Nous nous inspirons des écrits pertinents et de l'expérience d'International Alert dans des sociétés fragiles et touchées par des conflits pour souligner le rôle clé de la résilience. Pour renforcer la résilience, les acteurs des secteurs du climat, du développement, de la construction de la paix et gouvernementaux devraient surmonter les barrières bureaucratiques et institutionnelles et coopérer pour nouer des liens entre les systèmes thématiques et régionaux cloisonnés.

La comprensión respecto a qué involucra la resiliencia ante los cambios climáticos y ambientales exhibida por sociedades frágiles y afectadas por conflictos, reviste particular importancia y encierra retos específicos. Con el fin de explorar los vínculos clima-fragilidad-conflicto y clima-resiliencia-paz, los autores desarrollan un marco conceptual, a la vez que examinan distintos enfoques encaminados a establecer vasos comunicantes entre los conceptos de cambio climático y paz. En este sentido, y para destacar el importante rol desempeñado por la resiliencia, se apoyan en estudios previos y en la experiencia adquirida por International Alert en sociedades frágiles y afectadas por conflictos. A manera de conclusión, establecen que para fortalecer la resiliencia, las autoridades gubernamentales y los actores en los ámbitos de clima, de desarrollo y de construcción de la paz, por un lado, deberán superar las barreras burocráticas e institucionales y, por otro, cooperar de manera transversal en distintos temas y regiones.

Introduction

The relationship between climate vulnerability and conflict has received considerable attention from the academic and donor community (World Bank 2011; Gleditsch 2012; Scheffran et al. 2012a; USAID 2012). Some studies have identified an overlap of countries facing a double exposure of natural disasters and armed conflict (Scheffran et al. 2012c), whilst others attempt to demonstrate (Burke et al. 2009; Hsiang, Burke, and Miguel 2013) or reject (Buhaug 2010; Slettebak 2012) causal trends between climatic change and incidence of conflict. So far, the field is dominated by quantitative approaches analysing temperature and precipitation data in conjunction with conflict records (Scheffran et al. 2012c, 2012b; Theisen, Gleditsch, and Buhaug 2013 provide overviews).

These approaches are of limited use for the peacebuilding community, including governments and non-government organisations (NGOs), for two reasons. First, while quantitative approaches may show if there is a correlation between climate or weather and conflict variables, the approach provides no answers as to why there is a relationship or the nature of any statistical relationship. The approach therefore offers no diagnostics or potential entry points for peacebuilding which could influence or disrupt potential links between climate change and conflict. Second, in almost all statistical analyses, security is reduced to the absence or presence of armed conflict (e.g., Burke et al. 2009; Buhaug 2010; Slettebak 2012). This is in contrast to the security and development literature that has moved from focusing on state security (which might be measured in number of armed conflicts) to human security (Dalby 2009; Matthew et al. 2010). Parts of the policy-oriented literature offer a broader conceptualisation of security and climate along with a considerable degree of granularity on interacting factors and entry points for addressing them (WBGU 2007; Smith and Vivekananda 2009, 2007, 2012; Tänzler and Ries 2012).

Rigorous exploration of the relationship between climate and conflict on the one hand, and climate and peace on the other hand, is hindered not only by the complex interactions between these factors but also the complexity and ambiguity of the intermediate concepts. The concepts of fragility, vulnerability, adaptation, resilience, and human security have been discussed in other studies (e.g., Adger 2006; Duit et al. 2010; Brinkerhoff 2011) but rarely in conjunction (Smith and Vivekananda 2009; Scheffran, Link, and Schilling 2012d are exceptions).

Against this background, the present paper pursues two aims. First, we explore how the major concepts relevant for exploring the climate-fragility-conflict nexus relate to each other by developing a conceptual framework. Second, we discuss approaches to enable and promote pathways between climatic changes and peace in fragile and conflict-affected societies. Particular emphasis is placed on resilience and the institutional and structural changes to build it. The paper is structured along the two aims and a step-by-step development of the framework. We draw on the relevant literature of each research community and illustrate linkages using examples of International Alert's extensive experience and research in peacebuilding and climate vulnerability from fieldwork across Africa, South Asia, and Central Asia.

From climate change to conflict or peace

The negative cycle

Figure 1 shows a trajectory whereby fragility increases the vulnerability of communities to climate change. This vulnerability in turn can lead to human insecurity which potentially results in violent conflict between communities or communities and the government. Violent conflict can then further contribute to fragility, in a negative cycle. In the figure, a lighter arrow indicates an increasing relationship between factors. For example, when human insecurity increases the risk of violent conflict rises as well. A darker arrow indicates a decreasing impact. For

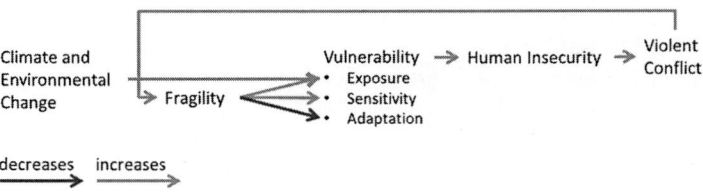

Figure 1.   The negative cycle.

example, a higher level of fragility is likely to decrease the capacity for adaptation. The linkages become clearer once we have a better understanding of all concepts involved.

The IPCC (Intergovernmental Panel on Climate Change) (2012, 557) defines climate change as "*a change in the state of the climate that can be identified (e.g., by using statistical tests) by changes in the mean and/or the variability of its properties and that persists for an extended period, typically decades or longer*". In contrast to the United Nations Framework Convention on Climate Change's (UNFCCC) definition (UNFCCC 1992), the IPCC does not distinguish between anthropogenic (human-caused) and natural climate change. Climatic changes are usually measured by changes in temperature and precipitation which affect humans indirectly through environmental changes. For example, an increase in water temperature in the Bay of Bengal leads to sea level rise which can then lead to floods. These are experienced by local communities as changes in their environment and to their livelihoods and well-being rather than global climate change (e.g., Schilling et al. 2013). Similarly, pastoralists in north-western Kenya notice a higher frequency of drought affecting the health of their herds rather than a 0.9°C temperature increase between 2006 and 2009 (Schilling et al. forthcoming). Some environmental changes may be the result of human actions, such as a flood caused by over-siltation of a river due to upstream deforestation. For affected communities in fragile contexts, the cause of the change in their environment does not have a bearing on the impact they experience. Hence, for the purposes of this paper, no causal distinction is made between climate change and environmental change and the two terms can be read and applied synonymously (for an application of environmental vulnerability see Schilling et al. 2013).

Most of the fragility literature has focused on weak governance and state fragility, which usually refers to states that are incapable (or unwilling) "*of assuring basic security, maintaining rule of law and justice, or providing basic services and economic opportunities for their citizens*" (GSDRC 2014, first paragraph; see also OECD 2008). In our framework fragility does not only apply to official government bodies but also to informal institutions, such as a council of elders. Examples of how governments have inadvertently increased the exposure and sensitivity to climatic changes include government incentives for communities to move from cultivating a diverse range of crops to mono-cropping. We found this to be the case in Rwanda. While our findings are not published yet, they confirm a study by Pritchard (2013, 186) who notes that "*the rapid and forceful implementation of tenure and agricultural policies is unnecessarily undermining the livelihood stability of rural subsistence farmers*". Similar processes are reported from Morocco where the government gives incentives for farmers to switch from barley to the more profitable but less drought resistant wheat (Schilling et al. 2012a). A weak government is also less capable of supporting adaptation to environmental stresses. For example, in the 2011 drought in the Horn of Africa, the Kenyan government was able to provide food relief to its population mostly without external support while in Somalia the government depended entirely on international aid (UNOCHA 2011a, 2011c, 2011b). In Nepal's Koshi basin, Vivekananda (2010) reports that the current context of political instability and criminal gangs hinder community-level adaptation to floods. Evans and Steven (2009, 2) argue that the policy challenge of climate change is above all "*one of leadership, co-ordination and collective action – and hence about institutions*". The German Advisory Council on Global Change (WBGU 2007, 54) argues that weak, conflict-affected states "*which even under current conditions barely have the capacity to maintain a functioning polity*" could be overwhelmed by the impact of climate change, "*potentially leading to distortions in international politics as well*". Several studies (e.g., Birch and Grahn 2007; Stark, Terasawa, and Ejigu 2011) report failures of governance that hamper efforts for both adaptation and conflict resolution. When communities perceive that their government has failed in supporting their resilience to climate and environmental risks, this failure can weaken the social contract between citizens and the state. For example, in Sindh, Pakistan, local communities expressed a negative perception towards the government whom they held responsible for biased distribution and control of land rights which posed an obstacle to adaptation to floods in 2010 and 2011 (Schilling et al. 2013). An example of how a weakening of informal institutions can decrease a community's capacity to mitigate violent conflict can be found in north-western Kenya. Here, the loss of respect for and influence of Turkana elders is reported to make conflict mitigation between the Turkana and other pastoral groups more challenging (Schilling, Opiyo, and Scheffran 2012b).

Vulnerability is usually divided into the three elements of exposure, sensitivity, and adaptation. Exposure refers to the rate and magnitude of change (for example temperature increase) that an area is experiencing (IPCC 2007). It is useful to interpret sensitivity with respect to its resource dimension as suggested and done by other authors (Barnett and Adger 2007; Schilling et al. 2012a). Sensitivity is then understood as a measurement of the availability of a resource (for instance water) prior to the climate change impact and its importance for the livelihoods in the affected area. For example, a farmer cultivating rain-fed crops in already arid Mali is more sensitive to a reduction in rainfall than a lawyer working in an office in London. The example shows that part of the vulnerability of many states in Africa is the strong dependence on rain-fed agriculture for employment and national income (Busby et al. 2012).

Adaptation is "*the process of adjustment to actual or expected climate and its effects, in order to moderate harm or exploit beneficial opportunities*" (IPCC 2012, 556). It is the element of vulnerability that is most actionable (see also Paavola and Adger 2006). While the richer parts of a society usually experience higher losses in monetary terms, they are also the ones having the assets to recover faster (for instance, because their property is insured). In contrast, poorer parts of the society are both more exposed to natural disaster (for example because they live in more basic shelters) and have fewer assets to recover. In understanding the social processes that turn an event into a disaster, Bankoff, Frerks, and Hilhorst (2004) emphasise the relative advantage or deprivation of a group within the social order. People in weaker social positions, meaning those with less financial resources, assets, knowledge, representation, and power, are in general more vulnerable to environmental and climatic changes.

Vulnerability can contribute to human insecurity at individual, household, and community level. Human insecurity is the opposite of human security which can be defined as the "*freedom from the risk of loss or damage to a thing that is important to survival and well-being*" (Barnett, Matthew, and O'Brian 2010, 4). The concept can be further specified into environmental security (the absence of risk or threat to the environment a person or community depends on and lives in), human well-being (the physical safety and security and psycho-social well-being), economic security (availability of economic resources, stability, institutions, and relations to provide for an adequate level of welfare), and food security (availability, stability, utilisation, and access to food) (King and Murray 2001; Khagram, Clark, and Raad 2003; Khagram and Ali 2006; Schmidhuber and Tubiello 2007; Dalby 2009; Barnett, Matthew, and O'Brian 2010). When the human security of people is negatively affected, people might resort to means and actions that are incompatible with other people pursuing their aims, especially when there are already other conflict drivers present (Bronkhorst 2011; Scheffran, Link, and Schilling 2012d; Bamidele 2013). While it must be acknowledged that conflict can act as a driver for positive change and development, violent conflict in which actors use force against other actors to pursue their goals, is widely seen as detrimental to human security (Collier et al. 2003; World Bank 2011; Gates et al. 2012).

An example of the full negative cycle exists in north-western Kenya. Increasing temperatures and rainfall variability decrease the availability and reliability of pasture and water for the environmentally sensitive pastoral communities. Kenya is not a fragile state, but in the north-west of the country, the government fails to cover the population's basic needs for security, infrastructure, and income opportunities. Instead, the mobility, as the key adaptation strategy for the pastoralists, is limited by the governments of Kenya and Uganda. The high vulnerability leads to human insecurity, specifically environmental and food insecurity. These insecurities serve as motivations for pastoral groups to use force against each other to gain or defend control over pasture and water. The resulting violent conflict in turn hinders development and government engagement and hence contributes further to fragility (Schilling 2012; Schilling, Opiyo, and Scheffran 2012b).

The positive cycle

Overall, Figure 1 shows the potential mechanisms that make fragile contexts more vulnerable to climate change and climate vulnerable societies facing greater risks of violent conflict. Figure 2 adds the opposite, positive cycle to this. The positive cycle reflects that stable societies, in which formal and informal institutions are able and willing to cover the population's basic needs, are more resilient and have a higher level of human security. This in turn makes peace more likely and stability easier to achieve. We define peace not only as the absence of violent conflict but rather as a situation in which people engage "*in inclusive social change processes that improve the quality of life*" (International Alert 2010, 15). As part of the discussion on how this "*positive peace*" (e.g., Amster 2013, 473) can be strengthened in response to climatic and environmental changes, the concept of resilience is increasingly receiving attention in the academic, NGO, and donor community (e.g., Mercy Corps 2012; USAID 2012; Hamza, Smith, and Vivekananda 2012; Chandler 2013). A growing body of empirical evidence suggests that communities do not face climate and environmental changes in isolation but rather in conjunction with socio-economic and political risks (Stark, Terasawa, and Ejigu 2011; Mercy Corps 2012; Schilling et al. 2013; Vivekananda et al. 2014). The concept of resilience lacks a universally accepted and precise definition. For the purposes of this study, resilience is defined as:

> "the ability of countries, communities, and households to anticipate, adapt to, and/or recover from the effects of potentially hazardous occurrences (natural disasters, economic instability, and conflict) in a manner that protects livelihoods, accelerates and sustains recovery, and supports economic and social development." (Inter-Agency Working Group on Resilience 2012, 7)

Several authors have emphasised the usefulness of resilience when approaching social-ecological systems (Duit et al. 2010; see also Gallopín 2006). However, as Adger (2006) has stressed

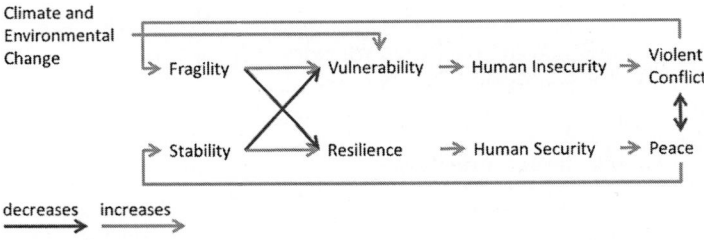

Figure 2.   The negative and positive cycles.

regarding vulnerability, it needs to be clear whose resilience and to what (see also Frazier et al. 2013). The OECD for example distinguishes between target groups (e.g., women), *"layers in society"* (families, communities, nations) (Mitchell 2013, 10), and *sector-specific systems* (e.g., food security) and thematic (e.g., poverty) units of analysis.

Resilience needs to be understood in response to multiple interacting hazards not to stand-alone single risks, and as such resilience to one risk can build resilience to others (Smith and Vivekananda 2009; Stark, Mataya, and Lubovich 2010; Tänzler, Maas, and Carius 2010; Mitchell 2013). Mazo (2010) sees strengthening adaptive capacity as supporting major public goods in governance, accountability, public health, and education. Goulden and Few (2011, 54) move in a similar direction when they report from their research on the Niger Basin that *"social, political and economic factors, such as past development policies, may have significantly influenced the vulnerability of some social groups"* to *"pressure on basic resources such as land, water and, therefore, food"*. From this they conclude that:

> "[B]uilding peace through resolving conflicts, adapting to the consequences of climate variability and climate change, and pursuing equitable and sustainable development are linked elements of enhancing human security and building resilience."

Likewise, a Mercy Corps (2012) study reports from southern Ethiopia that improving social cohesion and local institutions for conflict mitigation enhances access to natural resources, and that pastoralist groups with greater access recover more quickly from drought. In the Karamoja cattle corridor in Uganda, while there are many complexities in the local effects of government adaptation policies, Stark et al. found that local people:

> "were eager to adopt new techniques, take some risks, and share their knowledge and successes with others. This knowledge includes an already existing, sophisticated understanding of the environmental conditions in the Cattle Corridor. The engagement and use of indigenous knowledge is a key asset in addressing climate change impacts and enhancing community resilience in the Cattle Corridor." (Stark, Terasawa, and Ejigu 2011, 5)

Goulden and Few (2011, 3) conclude that peaceful adaptation depends on reasonably good governance, *"at least to ensure that state authorities are not an obstacle to adaptation"*. The examples show that governance, be it by governmental or non-governmental actors, can increase the resilience of communities which in turn can lead to positive linkages between human security and peace and again stability and good governance (Figure 2). The examples further illustrate the linked nature of community resilience, the benefits of the integrated approach of building resilience, and the social costs of not doing so. The following section discusses the policy implications of the negative and positive cycles.

Approaches to the climate-resilience-peace nexus

This section provides an overview of approaches for avoiding the negative cycle and for promoting the climate-resilience-peace nexus instead. As shown in Figure 3, we suggest that research, policy, and practice should focus on resilience while understanding the context specific complexity, revising the funding structure, and changing the institutions.

Understanding complexity to build resilience

In attempting to address the climate-resilience-peacebuilding nexus, the first point to note is that resilience is a complex phenomenon. As Evans (2010, 10) stresses, *"scarcity should not be viewed in isolation from the contextual factors that make an individual, community or society*

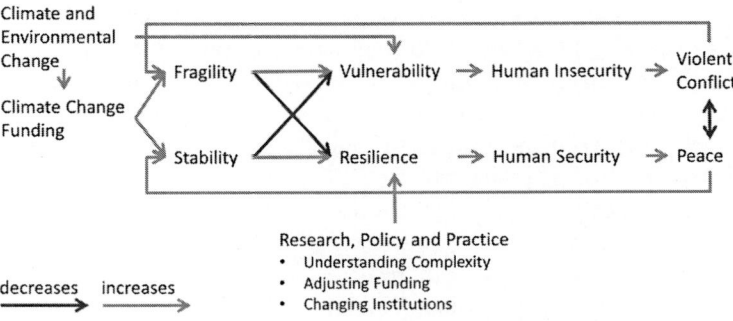

Figure 3.   Integrated framework of the climate change-conflict/peace nexus.

vulnerable – or resilient – to its effects". The case of the Semliki River, which forms part of the border between DRC and Uganda, illustrates the point well (Hamza, Smith, and Vivekananda 2012). Climatic changes influence flooding and the course of the river. Relations between the two states are further complicated by the recent discovery of oil and a history of conflict. In this and similar contexts, helping communities build resilience both requires state-society relations to be in good order on both sides of the Semliki River, and good state-to-state relations. Resilience is determined by decisions made at the household, national, regional, and international levels. Benjaminsen et al. (2012) conclude from researching land conflicts in the Sahel that climate variability is less of an issue than national political and economic circumstances, but also that in responding to drought, rice-farmers had moved downriver which might lead to confrontations, tensions, and conflict (see also Petrocelli, Newport, and Hamro-Drotz 2013). In other words, risks are ever-changing, affected not only by climatic changes and by impersonal forces such as economic trends or faraway government decisions, but also by decisions taken by those most closely involved. As Mitchell (2013, 11) notes, the risks people face are complex and interlinked: risks are "*typically made up of large disasters (intensive risk) interspersed with small frequent negative events (extensive risk), derived from a combination of natural, geo-political and economic factors, driven by a range of long-term and interconnected trends*". As such, efforts to research, analyse, and address resilience would benefit from taking account of the different target groups (e.g., women), layers in society (families, communities, nations) and sector-specific systems (e.g., food security) and thematic (e.g., poverty) unit of analysis and their interactions (Mitchell 2013).

Operationalising resilience-building into programming on the ground thus requires a deeper understanding of the risk landscape that people, institutions, and infrastructure face; the different layers of risks, and the interaction of risk factors across different layers. A good starting point may be to recall that in development aid, rule number one for operating in fragile states, according to the OECD (2011a, 23), is to "*take context as the starting point*". In the same vein, Stark, Mataya, and Lubovich (2009, 2) argue for "*more granularity in the understanding of the climate-conflict relationship*" in specific locales. Whilst necessary, community level context analysis is not sufficient. Community climate resilience in fragile contexts can be compromised where national or geopolitical risks are not taken into account. For example, efforts to build resilience to chronic food insecurity through food aid in Nepal inadvertently undermined long-term community resilience through creating a non-indigenous cultural dependency on rice in mountainous regions which lacked sufficient water to cultivate paddy (Schilling et al. 2013). Further, building resilience locally is important but it will be inadequate if the influence of national policies on local abilities to adapt is ignored.

The research community can contribute to knowledge on building resilience by conducting studies that combine and explore the relations between the concepts discussed in this paper. Of particular interest is the question of obstacles to building resilience that emerge in conflict-affected and fragile states. Here it is particularly appropriate to focus on local capacities and solutions, including their strengths and weaknesses, and what happens when they break down. Adaptation efforts should be analysed with respect to their implications for the relations within and between communities. The role, behaviour, and incentives for change of institutions at national and international levels and especially how formal and informal institutions relate to each other are key to understand resilience. With respect to climate, development, and peacebuilding funding, it is promising to deepen our understanding of elite behaviour and responses to incentives. Finally, more attention needs to be paid to the role of businesses and private security providers in fragile and conflict-affected states.

Climate change funding

Global aid related to climate change has the potential to contribute to fragility or stability (Figure 3). According to the UNFCCC (2007) the Green Climate Fund (GCF), set up to support developing countries in their efforts to mitigate and adapt to climate change, needs to reach between about US$49 billion and US$171 billion a year by 2030 for adaptation alone while estimates by Parry et al. (2009) go twice as high. The numbers show that the stakes are high. In theory, the resources from the GCF decrease the developing countries' vulnerability to climate change and hence contribute to their human security and a reduction of fragility (Figure 3). However, in most fragile states and many conflict-affected ones, power is organised so that dominant factions of the elite have privileged access to economic and political opportunity. This takes priority over the idea of society's leaders providing for the basic needs of the poorest citizens. Birch and Grahn (2007), for example, point out that pastoralists are in a weak position to benefit from adaptation funding as they are less politically connected and influential than other groups. Thus adaptation funding might not ease the climate and scarcity pressures faced by pastoralists.

Financing for resilience, be it through climate financing mechanisms, development aid, or other sources, needs to be long term and flexible, allowing for more joined-up programming and for greater adaptability of programmes to deal with the uncertainty and flux presented by climate change. Current climate financing mechanisms which require climate change spends to be new and additional to development aid, create structural barriers to joined-up approaches between climate change and development communities. On the ground, this leads to highly technical responses to climate change adaptation to ensure that climate change spends are not seen to overlap with development spends. Thus the funding architecture goes against any research and policy efforts to promote greater coherence which is critical to ensuring the conflict sensitivity of climate change aid. Risk management and resilience criteria should be part of all budget allocation processes. A *"tagging and tracing"* methodology would be a positive move away from silo budget lines and allow for greater integration of programming, whilst also enabling the tracking and calculation of allocated funds (Mitchell 2013). Aligning all financial flows to partner countries under a common risk framework could also be a valuable step to enhance aid effectiveness of climate aid in building resilience as per the renewed commitments of OECD member states at Busan (OECD 2011b).

Institutions

Formal and informal institutions play key roles in building resilience to climate change impacts and conflict. This section discusses institutional limitations, obstacles, and opportunities to build resilience.

Legitimacy of information and institutions

Legitimate, participatory, and transparent forms of governance are vital for reducing the risk of conflict and increasing the ability to adapt. Having adequate information in order to make informed decisions about the risks faced is a key element of community resilience (Mitchell 2013). However, access to climate data and information is not enough. Engagement within and between communities, government institutions, and other outside organisations is improved by better consultation when there is confidence in the accuracy of the information about problems and responses. Useful information therefore has at least four dimensions: relevance, accuracy, clarity, and legitimacy. Given the complexity of risk, useful information requires joint analysis of complex, changing, and uncertain risk landscapes. If one dimension is underdeveloped the information might not improve the resilience of a community. For example, the expert warnings on the Limpopo flooding of 2000 in Mozambique were rejected by village leaders because the community's traditional ways of understanding risk told them they were safe (Smith and Vivekananda 2009). The disregard of the warnings resulted in about 700 people drowning. Village institutions often seem to receive a degree of trust that is not given to local and national government (e.g., Schilling, Opiyo, and Scheffran 2012b; see also Weissmann 2012). Similarly, formal mechanisms of conflict management, such as the police and the courts seem less trusted than informal modes (for examples see Yami, Vogl, and Hauser 2011; Turner et al. 2012). There are however examples that show how formal institutions can support informal conflict resolution mechanisms. Goulden and Few (2011) report from Selingue in Mali that a scheme initiated by local government and supported by foreign assistance managed to reduce resource conflict between farmers and herders by promoting informal dialogue between the two groups. Others have stressed the potential of formal institutions to recognise and strengthen informal institutions as well (Vivekananda 2010; Stark, Terasawa, and Ejigu 2011; Mercy Corps 2012).

Working across silos

The combined task of responding to climate stresses and reducing the risk of conflict thus requires multiple scales of action – household, village, provincial, national, and international. It also necessitates working across sectoral silos. For the climate change community this means to ensure their adaptation efforts are peace-positive or at least conflict-sensitive. Babcicky (2013, 486) notes: "*conflict-sensitive adaptation understands the context it acts upon and strives to minimise negative and maximise positive impacts on human security.*"

This has two implications. First, moving away from the fixation on technical solutions to climate change, and second to broaden the scope from adaptation towards resilience. For the peacebuilding and development communities a shift towards resilience means to make their conflict mitigation and development efforts "*climate-proof*" (Smith and Vivekananda 2012). This in turn implies that it is not enough to consider conflict actors and development needs but also analyse how these are affected by and affect environmental changes. More helpful and harmonised engagement in resilience requires structural changes within donor and development agencies, strong governance, leadership, and adequate financing to tackle the complex root causes. This means recognising the synergy between development, adaptation, and peacebuilding. Practically, breaking down barriers between existing research and programming silos requires joint action across different research fields, development practitioners, and donors, emphasising complementary strengths, mandates, and interests to ensure a coherent approach. Shared risks analysis drawing on analytical frameworks from different academic strands would be a valuable contribution to resilience programming on the ground.

Incentives for institutional change

Joint action to promote complementary strengths, mandates, and interests across the academic, policy, and practitioner community working on resilience requires a number of steps to overcome entrenched disincentives and to create positive incentives for change. North, Wallis, and Weingast (2009) offer a comprehensive analysis of how privileges limit the use of violence by powerful individuals. Key questions which require further research around incentives include: How do you get traction? What are the incentives – especially for the elite – to deal with problems related to climate change in ways that are peaceful, law-based, inclusive, and recognisably democratic?

Starting with disincentives, these are largely structural, linked to the way aid is designed and operationalised. Donors, who shape the research agenda through research funding streams, have committed to aligning to partner country priorities, yet national governments in fragile contexts rarely prioritise risk management. This can also create a conflict of interest when national governments setting priorities comprise some of the risks communities face. Another disincentive is the lack of capacity within national governments of fragile states to deal with complex risk management. They may lack a centralised management structure or departmental home for the issue, adequate staff, absorptive capacity, and appropriate legal and regulatory structures. Cultural factors, economic difficulties, and insecurity which characterise fragile contexts can also disincentivise coordinated engagement. Within donor and development agencies, the separation of humanitarian and development programming is a challenge for an integrated approach to resilience.

Incentives for building resilience into existing and new research agendas include high-level political support, adequate resourcing for multi-disciplinary research and greater conceptual clarity and consensus around resilience. According to Mitchell (2013, 13), *"creating a rigorous technical approach for resilience programming in the field would be a good first step"*. Within the policy and practitioner community, efforts to embed risks and resilience across institutional structures and processes have started, though largely stopped with headquarters. Integration of resilience throughout organisations is critical not just in capitals but also in the field. Whilst technical guidance and leadership are necessary precursors to any incentivisation, change will require the creation of appropriate performance management incentives, creating a culture of peer contestability in programming, providing knowledge management systems and the promotion of the positive impacts of integrating resilience into programming, both locally and internationally (OECD 2012; Mitchell 2013).

International institutions

While some see a serious lack of capacity in international institutions to respond effectively (e.g., WBGU 2008), others point out that international institutions can and do work effectively. Tir and Stinnett (2012) for example stress the importance of institutions to mitigate conflicts over international water bodies which are affected by climate change.

In both the Nile Basin and the Niger Basin, the value of agreements and their institutionalisation has been extolled as means of conflict mitigation (Kameri-Mbote and Kindiki 2009; Goulden and Few 2011). This reflects a more general case for seeing environmental cooperation as a means of building peaceful relations. In the Great Lakes region in Central Africa, Kameri-Mbote (2006) points out that sharing environmental dependencies across state borders establishes a unifying force, facilitating dialogue and encouraging collaboration. Similar arguments have been drawn from both Central and South Asia (Conca and Dabelko 2002; Conca 2007) and are made more broadly in various case studies and general surveys (e.g., Smith and Vivekananda 2009, 2007; Stark, Mataya, and Lubovich 2009; UNEP 2009).

While this is the promise, it is not an easy win. There is a growing literature on the difficulty of getting international institutions to change, even when they have formally subscribed to these changes. Bell (2008) explores the World Bank's attempts to make its work more conflict sensitive and shows that while power relations and identities such as caste or ethnicity have widely acknowledged impact on governance and access to economic opportunity, the World Bank is institutionally predisposed to downplaying the importance of these dynamics and the effects it has on them. Similarly Batmanglich and Stephen (2011) find limited support within the World Bank system for initiatives designed to strengthen the institution's impact in complex and fragile environments. The authors trace this in part back to the daily details of how institutions work, how job responsibilities are defined, how time is managed, what tasks are prioritised, and what successes acknowledged. Smith and Vivekananda (2012) provide further discussion on what is needed in getting the institutions right to address the climate-conflict-fragility nexus.

Conclusions

The focus of the quantitative literature on identifying correlations and arguing causality between climate change and conflict has been of limited value for the peacebuilding community as it provides no answer to the question of how climatic changes and conflict might be related. The framework set out in this paper stresses the need to understand the linked conceptual pairs of fragility and stability, vulnerability and resilience, and human security and insecurity, in order to analyse the pathways between climate change and violent conflict or peace. It has been previously acknowledged that factors such as instability, inequality, and poverty making a society vulnerable to conflict are the same drivers that make a society vulnerable to climate and environmental changes. On the one hand, this means that there is potential for a vicious cycle connecting climate change, vulnerability, and violent conflict. On the other hand, this implies that there is a positive cycle between climate change, resilience, and peace. The double dividend of resilience to conflict and climate change can only be achieved if the contextual complexities are taken into account. For the climate change community this means to ensure their adaptation efforts are peace-positive, and for the peacebuilding and development communities it means to ensure their conflict mitigation and development efforts are climate-proof. While in theory this integrated, resilience-centred approach is highly promising, its practical implementation requires far-reaching structural changes. Climate, development, peacebuilding, and government actors (including donors, NGOs, governments, and practitioners) would have to overcome bureaucratic and institutional barriers and cooperate across thematic and regional silos.

Acknowledgements
We thank the anonymous reviewer for the constructive comments.

References

Adger, W. N. 2006. "Vulnerability." *Global Environmental Change* 16 (3): 268–281.

Amster, R. 2013. "Toward a Climate of Peace." *Peace Review* 25 (4): 473–479.

Babcicky, P. 2013. "A Conflict-sensitive Approach to Climate Change Adaptation." *Peace Review* 25 (4): 480–488.

Bamidele, O. 2013. "Climate Change, War, and Global Struggle." *Peace Review* 25 (4): 510–517.

Bankoff, G., G. Frerks, and D. Hilhorst. 2004. *Mapping Vulnerability: Disasters, Development and People*. London: Earthscan.

Barnett, J., and W. N. Adger. 2007. "Climate Change, Human Security and Violent Conflict." *Political Geography* 26: 639–655.

Barnett, J., R. Matthew, and K. O'Brian. 2010. "Global Environmental Change and Human Security: An Introduction." In *Global Environmental Change and Human Security*, edited by R. Matthew, J. Barnett, B. McDonald, and K. O'Brian, 1–32. Cambridge: MIT Press.

Batmanglich, S., and M. Stephen. 2011. *Peacebuilding, the World Bank and the United Nations*. London: International Alert.

Bell, E. 2008. *The World Bank in Fragile and Conflict-affected Countries: 'How', not 'How Much'*. London: International Alert.

Benjaminsen, T. A., K. Alinon, H. Buhaug, and J. T. Buseth. 2012. "Does Climate Change Drive Land-use Conflicts in the Sahel?." *Journal of Peace Research* 49 (1): 97–111.

Birch, I., and R. Grahn. 2007. *Pastoralism – Managing Multiple Stressors and the Threat of Climate Variability and Change*. New York: UNDP.

Brinkerhoff, D. W. 2011. "State Fragility and Governance: Conflict Mitigation and Subnational Perspectives." *Development Policy Review* 29 (2): 131–153.

Bronkhorst, S. 2011. *Climate Change and Conflict: Lessons for Conflict Resolution from the Southern Sahel of Sudan*. Durban: Accord.

Buhaug, H. 2010. "Climate not to Blame for African Civil Wars." *Proceedings of the National Academy of Sciences* 107 (38): 16477–16482.

Burke, M. B., E. Miguel, S. Satyanath, J. A. Dykema, and D. B. Lobell. 2009. "Warming Increases the Risk of Civil War in Africa." *Proceedings of the National Academy of Sciences* 106 (49): 20670–20674.

Busby, J., T. Smith, K. White, and S. Strange. 2012. "Locating Climate Insecurity: Where are the Most Vulnerable Places in Africa?." In *Climate Change, Human Security and Violent Conflict*, edited by J. Scheffran, M. Brzoska, H. G. Brauch, P. M. Link, and J. Schilling, 463–511. Berlin: Springer.

Chandler, D. 2013. "International Statebuilding and the Ideology of Resilience." *Politics* 33 (4): 276–286.

Collier, P., V. L. Elliott, H. Hegre, A. Hoeffler, M. Reynal-Querol, and N. Sambanis. 2003. *Breaking the Conflict Trap: Civil War and Development Policy*. Oxford: Oxford University Press.

Conca, K. 2007. "Transnational Dimensions of Freshwater Ecosystem Governance." In *Governance as a Trialogue: Government-society-science in Transition* by Johanna Hattingh, 101–22. Brelin: Springer.

Conca, K., and G. Dabelko, eds. 2002. *Environmental Peacemaking*. Washington, DC: Woodrow Wilson Center.

Dalby, S. 2009. *Security and Environmental Change*. Cambridge: Polity.

Duit, A., V. Galaz, K. Eckerberg, and J. Ebbesson. 2010. "Governance, Complexity, and Resilience." *Global Environmental Change* 20 (3): 363–368.

Evans, A. 2010. "Resource Scarcity, Climate Change and the Risk of Violent Conflict." *World Development Report 2011 – Background Paper*. New York: New York University/Center on International Cooperation.

Evans, A., and D. Steven. 2009. *An Institutional Architecture for Climate Change: A Concept Paper.* New York: Center on International Cooperation.

Frazier, T. G., C. M. Thompson, R. J. Dezzani, and D. Butsick. 2013. "Spatial and Temporal Quantification of Resilience at the Community Scale." *Applied Geography* 42: 95–107.

Gallopín, G. C. 2006. "Linkages between Vulnerability, Resilience, and Adaptive Capacity." *Global Environmental Change* 16 (3): 293–303.

Gates, S., H. Hegre, H. M. Nygard, and H. Strand. 2012. "Development Consequences of Armed Conflict." *World Development* 40 (9): 1713–1722.

Gleditsch, N. P., ed. 2012. "Special Issue on Climate Change and Conflict." *Journal of Peace Research* 49 (163): 1–145.

Goulden, M., and R. Few. 2011. *Climate Change, Water and Conflict in the Niger River Basin.* London: International Alert.

GSDRC. 2014. "Definitions and Typologies of Fragile States." Accessed January 23, 2014. http://www.gsdrc.org/go/fragile-states/chapter-1-understanding-fragile-states/definitions-and-typologies-of-fragile-states

Hamza, M., D. Smith, and J. Vivekananda. 2012. *Difficult Environments: Bridging Concepts and Practice for Low Carbon Climate Resilient Development.* Brighton: Institute of Development Studies.

Hsiang, S. M., M. Burke, and E. Miguel. 2013. "Quantifying the Influence of Climate on Human Conflict." *Science* 341 (6151): 1189–1213.

Inter-Agency Working Group on Resilience. 2012. "The Characteristics of Resilience Building." Accessed April 2, 2013. http://community.eldis.org/?233@@.5ad4406d!enclosure=.5ad4406e&ad=1

International Alert. 2010. *Programming framework for international alert.* London: International Alert.

IPCC. 2007. *Climate change 2007. Climate Change Impacts, Adaptation and Vulnerability.* Geneva: Cambridge University Press.

IPCC. 2012. *Managing the Risks of Extreme Events and Disasters to Advance Climate Change Adaptation.* Cambridge: Cambridge University Press.

Kameri-Mbote, P. 2006. *Conflict and Cooperation: Making the Case for Environmental Pathways to Peacebuilding in the Great Lakes Region.* Washington: Woodrow Wilson International Center for Scholars.

Kameri-Mbote, P., and K. Kindiki. 2009. "Water and Food Security in the Nile River Basin: Perspectives of Governments and NGOs of Upstream Countries." In *Facing Global Environmental Change*, edited by H. G. Afes-Press, N. C. Behera, P. Kameri-Mbote, J. Grin, U. Oswald Spring, B. Chourou, C. Mesjasz, and H. Krummenacher, 651–659. Berlin: Springer.

Khagram, S., and S. Ali. 2006. "Environment and Security." *Annual Review of Environment and Resources* 31 (1): 395–411.

Khagram, S., W. Clark, and D. F. Raad. 2003. "From the Environment and Human Security to Sustainable Security and Development." *Journal of Human Development* 4 (2): 289–313.

King, G., and C. J. L. Murray. 2001. "Rethinking Human Security." *Political Science Quarterly,* 116 (4): 585–610.

Matthew, R., J. Barnett, B. McDonald, and K. O'Brian, eds. 2010. *Global Environmental Change and Human Security.* Cambridge, MA: MIT Press.

Mazo, J. 2010. *Climate Conflict: How Global Warming Threatens Security and what to do about it.* London: Routledge.

Mercy Corps. 2012. *From Conflict to Coping: Evidence from Southern Ethiopia on the Contributions of Peacebuilding to Drought Resilience among Pastoralist Groups.* Portland: Mercy Corps.

Mitchell, A. 2013. *Risk and Resilience – From Good Idea to Good Practice.* Paris: OECD.

North, D., J. Wallis, and B. Weingast. 2009. *Violence and Social Orders: A Conceptual Framework for Interpreting Recorded Human History.* Cambridge: Cambridge University Press.

OECD. 2008. *Concepts and Dilemmas of State Building in Fragile Situations.* Paris: OECD.

OECD. 2011a. *International Engagement in Fragile States – Can't we do Better?* Paris: OECD.

OECD. 2011b. "OECD and the Fourth High Level Forum on Aid Effectiveness." Accessed January 31, 2014. http://www.oecd.org/dac/effectiveness/oecdandthefourthhighlevelforumonaideffectiveness.htm

OECD. 2012. "Analyses of Aid – How much Aid is Delivered, where and for what Purpose?." Accessed February 8, 2012. http://www.oecd.org/document/48/0,3746,en_2649_34447_42396656_1_1_1_1,00.html

Paavola, J., and W. N. Adger. 2006. "Fair Adaptation to Climate Change." *Ecological Economics* 56 (4): 594–609.

Parry, M., N. Arnell, P. Berry, D. Dodman, S. Fankhauser, C. Hope, et al. 2009. *Assessing the Costs of Adaptation to Climate Change: A Review of the UNFCCC and other Recent Estimates.* London: International Institute for Environment and Development & Grantham Institute for Climate Change.

Petrocelli, T., S. Newport, and D. Hamro-Drotz. 2013. "Climate Change and Peacebuilding in the Sahel." *Peace Review* 25 (4): 546–551.

Pritchard, M. F. 2013. "Land, Power and Peace: Tenure Formalization, Agricultural Reform, and Livelihood Insecurity in Rural Rwanda." *Land Use Policy* 30 (1): 186–196.

Scheffran, J., M. Brzoska, H. G. Brauch, P. M. Link, and J. Schilling, eds. 2012a. *Climate Change, Human Security and Violent Conflict: Challenges for Societal Stability.* Berlin: Springer.

Scheffran, J., M. Brzoska, J. Kominek, P. M. Link, and J. Schilling. 2012b. "Climate Change and Violent Conflict." *Science* 336 (6083): 869–871.

Scheffran, J., M. Brzoska, J. Kominek, P. M. Link, and J. Schilling. 2012c. "Disentangling the Climate-conflict Nexus: Empirical and Theoretical Assessment of Vulnerabilities and Pathways." *Review of European Studies* 4 (5): 1–13.

Scheffran, J., P. M. Link, and J. Schilling. 2012d. "Theories and Models of Climate-security Interaction: Framework and Application to a Climate Hot Spot in North Africa." In *Climate Change, Human Security and Violent Conflict: Challenges for Societal Stability*, edited by J. Scheffran, M. Brzoska, H. G. Brauch, P. M. Link, and J. Schilling, 91–131. Berlin: Springer.

Schilling, J. 2012. *On Rains, Raids and Relations: A Multimethod Approach to Climate Change, Vulnerability, Adaptation and Violent Conflict in Northern Africa and Kenya.* Hamburg: University of Hamburg.

Schilling, J., M. Akuno, J. Scheffran, and T. Weinzierl. Forthcoming. "On Raids and Relations: Climate Change, Pastoral Conflict and Adaptation in Northwestern Kenya." In *Climate Change and Conflict: Where to for Conflict Sensitive Climate Adaptation in Africa?*, edited by S. Bronkhorst, and U. Bob. Berlin: Springer.

Schilling, J., K. P. Freier, E. Hertig, and J. Scheffran. 2012a. "Climate Change, Vulnerability and Adaptation in North Africa with Focus on Morocco." *Agriculture, Ecosystems & Environment* 156 (0): 12–26.

Schilling, J., F. Opiyo, and J. Scheffran. 2012b. "Raiding Pastoral Livelihoods: Motives and Effects of Violent Conflict in North-western Kenya." *Pastoralism* 2 (25): 1–16.

Schilling, J., J. Vivekananda, P. Nisha, and M. A. Khan. 2013. "Vulnerability to Environmental Risks and Effects on Community Resilience in Mid-west Nepal and South-east Pakistan." *Environment and Natural Resources Research* 3 (4): 1–19.

Schmidhuber, J., and F. Tubiello. 2007. "Global Food Security Under Climate Change." *Proceedings of the National Academy of Sciences of the United States of America* 104 (50): 19703–8.

Slettebak, R. T. 2012. "Don't Blame the Weather! Climate-related Natural Disasters and Civil Conflict." *Journal of Peace Research* 49 (1): 163–176.

Smith, D., and J. Vivekananda. 2007. *A Climate of Conflict – The Links between Climate Change, Peace and War.* London: International Alert.

Smith, D., and J. Vivekananda. 2009. *Climate Change, Conflict and Fragility.* London: International Alert.

Smith, D., and J. Vivekananda. 2012. "Climate Change, Conflict, and Fragility: Getting the Institutions Right." In *Climate Change, Human Security and Violent Conflict*, edited by J. Scheffran, M. Brzoska, H. G. Brauch, P. M. Link, and J. Schilling, 77–90. Berlin: Springer.

Stark, J., C. Mataya, and K. Lubovich. 2009. *Climate Change, Adaptation, and Conflict - A Preliminary Review of the Issues.* Washington, DC: USAID.

Stark, J., C. Mataya, and K. Lubovich. 2010. *Energy Security and Conflict: A Country-level Review of the Issues.* Washington, DC: Foundation for Environmental Security and Sustainability.

Stark, J., K. Terasawa, and M. Ejigu. 2011. "Climate change and conflict in pastoralist regions of Ethiopia: Mounting Challenges, Emerging Responses." Accessed July 4, 2013. http://inec.usip.org/resource/climate-change-and-conflict-pastoralist-regions-ethiopia-mounting-challenges-emerging-respo

Tänzler, D., and F. Ries. 2012. "International Climate Change Policies: The Potential Relevance of redd+ for Peace and Stability." In *Climate Change, Human Security and Violent Conflict: Challenges for Societal Stability*, edited by J. Scheffran, M. Brzoska, H. G. Brauch, P. M. Link, and J. Schilling, 695–706. Berlin: Springer.

Tänzler, D., A. Maas, and A. Carius. 2010. "Climate Change Adaptation and Peace." *Wiley Interdisciplinary Reviews: Climate Change* 1 (5): 741–750.

Theisen, O. M., N. P. Gleditsch, and H. Buhaug. 2013. "Is Climate Change a Driver of Armed Conflict?" *Climatic Change* 117 (3): 613–625.

Tir, J., and D. M. Stinnett. 2012. "Weathering Climate Change: Can Institutions Mitigate International Water Conflict?" *Journal of Peace Research* 49 (1): 211–225.

Turner, M. D., A. A. Ayantunde, K. P. Patterson, and E. D. Patterson. 2012. "Conflict Management, Decentralization and Agropastoralism in Dry Land West Africa." *World Development* 40 (4): 745–757.

UNEP. 2009. *From Conflict to Peacebuilding: The Role of Natural Resources and the Environment.* Nairobi: UNEP.

UNFCCC. 1992. *United Nations Framework Convention on Climate Change.* New York: the United Nations.

UNFCCC. 2007. *Investment and Financial Flows to Address Climate Change.* Bonn: United Nations Framework Convention on Climate Change.

UNOCHA. 2011a. *Humanitarian Report Eastern Africa* (Vol. 3). Nairobi: UNOCHA.

UNOCHA. 2011b. "Kenya Emergency Humanitarian Response Plan 2011." Accessed August 18, 2011. http://ochadms.unog.ch/quickplace/cap/main.nsf/h_Index/MYR_2011_Kenya_EHRP/$FILE/MYR_2011_Kenya_EHRP_SCREEN.pdf?openElement

UNOCHA. 2011c. "Somalia: Extreme Concern Over the Deteriorating Drought Situation." Accessed August 18, 2011. http://www.unocha.org/top-stories/all-stories/somalia-extreme-concern-over-deteriorating-drought-situation

USAID. 2012. *USAID Climate and Development Strategy.* Washington, DC: USAID.

Vivekananda, J. 2010. *Climate Change, Governance and Fragility – Rethinking Adaptation.* London: International Alert.

Vivekananda, J., J. Schilling, S. Mitra, and N. Pandey. 2014. "On Shrimp, Salt and Security: Livelihood Risks and Responses in South Bangladesh and East India." *Environment, Development and Sustainability* January 2014: 1–21. http://link.springer.com/article/10.1007%2Fs10668-014-9517-x

WBGU. 2007. "World in Transition – Climate Change as a Security Risk." Accessed December 12, 2010. http://www.wbgu.de/wbgu_jg2007_engl.html

WBGU. 2008. *World in Transition – Climate Change as a Security Risk.* London: Earthscan.

Weissmann, M. 2012. *East Asian Peace: Conflict Prevention and Informal Peacebuilding.* Basingstoke: Palgrave.

World Bank. 2011. *World Development Report 2011: Conflict, Security and Development.* Washington, DC: World Bank.

Yami, M., C. Vogl, and M. Hauser. 2011. "Informal Institutions as Mechanisms to Address Challenges in Communal Grazing Land Management in Tigray, Ethiopia." *International Journal of Sustainable Development and World Ecology* 18 (1): 78–87.

Lessons from urban risk assessments in Latin American and Caribbean cities

Robin Bloch, Nikolaos Papachristodoulou, Rawlings Miller, Jose Monroy, Tiguist Fisseha, Lorena Trejos, Melanie S. Kappes, and Beatriz Pozueta

This paper draws on the results from a recent World Bank-funded project designed to inform policy-making and climate change adaptation planning in small and medium-sized cities in Latin America and the Caribbean. The focus was on floods and landslides, which are the two most common climate-related risks in cities across the region. The project allowed the application of the Urban Risk Assessment (URA) tool developed by the World Bank and the drawing of valuable lessons which may also be applicable to the many methods and tools available for climate change adaptation planning in the rapidly urbanising cities of developing countries.

Cet article se base sur les résultats d'un récent projet financé par la Banque mondiale conçu pour éclairer la formulation de politiques et la planification de l'adaptation au changement climatique dans des villes petites et moyennes d'Amérique latine et des Caraïbes. L'axe central tournait autour des inondations et des glissements de terrain, qui sont les deux risques liés au climat les plus communs parmi les villes de la région. Ce projet a permis d'appliquer l'outil d'évaluation des risques en milieu urbain (*Urban Risk Assessment* – URA), conçu par la Banque mondiale, et de tirer des enseignements précieux qui seront peut-être aussi applicables aux nombreux outils et méthodes disponibles pour la planification de l'adaptation au changement climatique dans les villes en voie d'urbanisation rapide des pays en développement.

El presente artículo se apoya en los resultados obtenidos por un proyecto recientemente financiado por el Banco Mundial, el cual se encaminó a proveer información que posibilitara el diseño de políticas y la planeación de la adaptación al cambio climático en ciudades pequeñas y medianas de América Latina y el Caribe. Con este objetivo, el enfoque del mismo se centró en los riesgos más comúnmente relacionados con el cambio climático en las ciudades de dicha región, es decir, en las inundaciones y los derrumbes. Durante el transcurso del proyecto, se aplicó la Evaluación del Riesgo Urbano (ERU), herramienta desarrollada por el Banco Mundial. Los valiosos aprendizajes surgidos de esta experiencia podrían ser aplicados en los múltiples métodos y herramientas existentes para la planificación de la adaptación al cambio climático en las ciudades de rápida urbanización de los países en desarrollo.

Introduction

Latin America and the Caribbean's (LAC) current urbanisation level is almost 80%, and the continent is experiencing a decelerating rate of urban population growth (UN 2012). This growth is becoming concentrated in cities of less than one million dwellers, where most of the urban population in Latin America and the Caribbean currently lives (UN-HABITAT 2012). Despite this slowdown, the built-up area of most cities continues to expand at rates that can be two or three times more than the rate of population increase (UN-HABITAT 2012). New urban development occurs horizontally (peripheral expansion) and/or "vertically" (often up slopes). According to Seto, Güneralp, and Hutyra (2012), forecasts of urban expansion in South America, based on the highest probability, show an increase of 67% between 2000 and 2030.

Urbanisation is thus accompanied by spatial expansion as polycentric structures appear, marked by new economic centres and a new juxtaposition of residential zones containing both prosperity and poverty. At the same time, urbanisation drives local and regional environmental changes, by altering land cover, hydrological systems, and biogeochemistry (Seto, Sánchez-Rodríguez, and Fragkias 2010).

Spatial expansion often occurs in hazard-prone areas, such as floodplains in both coastal and inland areas, as well as on landslide-prone slopes. Consequently, the exposure of urban populations to natural hazards is increasing. According to the 2011 United Nations Global Assessment Report on Disaster Risk Reduction, urban areas in Latin America concentrate more than 80% of all reported disasters (UNISDR 2011). Disasters appear to have been increasing faster in small and medium-sized cities (UNISDR 2011). Flooding is the most frequent hazard (Jha et al. 2011). Landslides

Table 1. Results from the world bank-funded project, "Climate change adaptation planning in Latin American and Caribbean cities."

The World Bank's Climate Change Adaptation Planning in Latin American and Caribbean Cities initiative started in April 2010 and will be completed in 2013. It consists of three phases.

The first phase was carried out by consultants based in each city and involved an initial institutional mapping and rapid diagnostic for the initiative. An ICF International consortium was commissioned in May 2012 to carry out second phase activities for the initiative. For each city, there were four main activities for this second phase: (1) a climate-related risks assessment focused on floods and landslides (2) a socio-economic adaptive capacity assessment (3) an institutional adaptive capacity assessment, and (4) based on the findings of the three assessments, a combined strategic climate adaptation institutional strengthening and investment plan, which will complement and be integrated into existing urban, environmental, and disaster risk reduction planning instruments for each city.

The outputs from the above-mentioned activities constitute a critical input for the final output of the overall initiative in its third phase: a regional Guidebook for city stakeholders on urban adaptation to climate change. Some brief context for each city is as follows:

Castries is the capital of Saint Lucia and the largest urban centre on the island, with approximately 65,000 inhabitants. Castries is susceptible to a multitude of natural hazards such as storm surges, hurricanes, tropical cyclones, landslides, and other extreme weather events.

Cusco sits high in the Andes Mountains in a valley within the upper basin of the Huatanay River at 3400 m above sea level. Cusco is exposed to a wide variety of natural disasters, including landslides and floods. The city of Cusco has a population of 400,000 inhabitants.

El Progreso with a population of approximately 200,000 inhabitants is located in the Sula Valley in North-western Honduras. It is located at the crossroads of two of the most important highways of the country. Flooding is a long-run problem for El Progreso and for the Sula Valley.

Estelí is located in North-western Nicaragua. It has a population of approximately 100,000 inhabitants. The Estelí River and its two tributaries, which run through the city, are responsible for the floods occurring in the city.

Santos is located on the southern Brazilian Coast, in the estuarine system of Santos. It has a population of approximately 420,000 inhabitants. The city is susceptible to both flooding and landslides.

---

triggered by rainfall and storm events are also common. Indeed, current urban growth patterns appear to have significantly amplified the exposure of urban populations to hazard risks, markedly but not exclusively those broadly characterised as the urban poor (Bloch, Papachristodoulou, and Brown 2013). Climate change is likely to worsen their impacts in the future.

In this paper we discuss the application of urban risk assessments in small and medium-sized cities in Latin America and the Caribbean, and highlight the need to link or incorporate climate change adaptation into existing priorities, sectoral plans, and planning instruments. Our discussion draws on the results from a recently conducted World Bank-funded project "Climate Change Adaptation Planning in Latin American and Caribbean Cities" (Table 1). The project was designed to inform policy-making and climate change adaptation planning in five small and medium-sized cities in the region: Castries, Saint Lucia; Cusco, Peru; El Progreso, Honduras; Estelí, Nicaragua; and Santos, Brazil.

The primary focus was cities less likely to have had access to climate change adaptation training, finance, or knowledge networks. The emphasis was on floods and landslides, which are two of the most common climate-related risks in cities across the LAC region. Poorly planned and managed urban development and spatial expansion also contribute to flood and landslide hazard risks. In assessing how climate change may have an effect on future flooding and landslide impacts and vulnerabilities, we attempted to gain insight into how climate-related hazards currently interrelate with trends in urban and economic development and institutional change.

The URA tool

The Terms of Reference for the activities carried out under this project derived primarily from an Urban Risk Assessment (URA) tool developed by the World Bank. The tool's assessment methodology *"focuses on three reinforcing pillars that collectively contribute to the understanding of urban risk"* (World Bank 2011, 2): a hazard impact assessment, an institutional assessment, and a socio-economic assessment. Assessments in the URA tool are associated with three levels of complexity (primary, secondary, and tertiary). The primary level provides an entry point to assess the challenges posed by climate-related hazards. The secondary level provides a more refined analysis to identify and map the most vulnerable areas and populations exposed to climate-related hazards and to consider how hazards may change in the future. Finally, the tertiary level undertakes specific probabilistic risk assessments and makes use of advanced risk management tools.

Progression from the primary to the tertiary level in any city or town is dependent upon the availability of what can be (and often is required to be) significant amounts of data, the technical capabilities of relevant staff and actors, and the ability and willingness of politicians, officials, and others to commit what can amount to not inconsiderable financial resources and time for conducting assessments – and to formulating and implementing policy, strategy, and action plans on the basis of findings.

Using the URA for assessments in LAC cities

The URA is avowedly a flexible tool, as it needs to be. In all five cities, however, data availability and time and resource constraints meant the following adaptations to the URA approach.

Climate-related risks assessment for floods and landslides

This analysis utilised existing planning tools, data, and resources used by the cities to consider how flood and landslide hazards may change by approximately the mid twenty-first century.

To effectively inform future disaster risk and urban planning, it was important that the approach be appropriately aligned with the available local data and planning procedures.

It was possible to assess present-day current flood and landslide **hazard levels**, which are studied and well-understood in all the cities. However, the city-scale information available across the cities varied in temporal and spatial resolutions, providing an opportunity to conduct assessments at varying scales of available information. To investigate future changes in floods and landslides, three approaches were considered and the viability of implementing each one based on available information in the cities was tested:

- The first approach identified and interrogated the flood and landslide hazard maps used by local stakeholders in planning and emergency management. Any precipitation metrics used to develop the flood and landslide maps were identified. An analysis was done to quantify how these precipitation metrics may change in the future and a discussion of the implications of these changes on the frequency and/or intensity of future flood and landslide events was provided.
- Using regional meteorological events that have caused floods and/or landslides was also a useful approach in developing precipitation event thresholds. How floods and/or landslides may change in the future was then investigated by examining future daily precipitation projections to see how often these thresholds might be crossed in the future.
- When observational data and/or records are very limited, global datasets of precipitation projections gave insight to how changes in the nature of precipitation may impact future floods and landslides. Considering how precipitation indicators such as the five-day maximum per year, the 95-percentile, the number of days above 10 millimetres of rain per year, may change provided some indication of how the frequency, duration, and intensity of events may change on a daily scale. Though these indicators are not always developed based on identified site-specific metrics, they do provide information regarding future changes in storm events that can be useful when considering how climate change may affect hazard events.

The level of detail in the findings for use by the cities diminishes from the first approach to the third approach, moving from a more quantitative analysis to one that is more qualitative. To investigate how landslides and floods may change in the future in Castries, for example, we largely adopted the first approach. For Santos, we mostly adopted the first approach supplemented with additional information from the third approach to investigate how the occurrence and intensity of landslides and floods may change in the future.

Given the constraints on the available information, we also adopted the third approach to investigate how landslides and floods may change in the future for Cusco, El Progreso, and Esteli. Future work should develop flood and landslide precipitation indicators based on today's relationships. The projections of these indicators would provide some insight regarding potential changes in floods and landslides. However, the lack of historical and existing meteorological observations is an immediate challenge for these cities to conduct such an analysis.

For all cities in the project, the uncertainty in precipitation projections and other non-climate factors that affect flood and landslide risks needs to be considered in the application of these findings by decision makers. Linking the potential climate projections to the way urban development is taking place is essential in understanding the possible effects that climate change could have in the cities. For example, although climate projections may show a potential decrease in seasonal precipitation and a rise in seasonal temperature by mid-century, as the finding for El Progreso and Esteli demonstrated (Table 2), which might result in a decrease in flood and landslide risk

Table 2.   Projected climate changes by city.

	Projected climate changes
Castries	Climate projections suggest that inland flood and landslide risk may be reduced as the intensity of rainfall patterns is generally expected to decrease. On the other hand, coastal flooding may be increased as impact of storm surge and waves will intensify with rising sea level.
Cusco	The rainy season may be intensified and extended by a few months, which would slightly increase precipitation. This would also result in an increased possibility of floods and landslides. However, projections also suggest that there could be a reduction in the intensity of rainfall during precipitation events: i.e., there could be more frequent, yet less severe rainfall events, which might ultimately moderate the possibility of floods and landslides. The aforementioned uncertainty is further complicated by Cusco's variable terrain, which can lead to large differences in precipitation received in one location versus another.
El Progreso	Temperatures are projected to increase. The threat of floods and landslides may be reduced as the seasonal and annual rainfall is generally expected to decrease (though it is not clear how the intensity, frequency, and duration of rainfall events may change).
Estelí	Both the wet and dry seasons are projected to become drier. Overall, the projected reduction in precipitation may decrease the possibility of flooding. The reduced soil moisture due to precipitation decreases may reduce the threat of landslides for unaltered vegetation. However, drying soils may increase the threat of landslides where vegetation has been removed and/or covered by impermeable surfaces.
Santos	Precipitation patterns are projected to change in the future, possibly resulting in a decrease in precipitation-induced floods, though coastal flooding from storm surge and high tide may increase. There may be a decrease in the overall frequency of landslides, but an increase in the occurrence of extreme landslide events.

assuming precipitation events that may trigger these hazards do not become more severe, the trends in urban development could actually lead to risk remaining constant or even increasing.

Full assessment of the **risk levels** for the flooding and landslide hazards, both currently and for the future, was not possible in any of the cities as the financial and demographic data necessary was not readily available. The climate-related risks assessment was therefore titled – and more correctly seen as – a climate-related hazard assessment. In addition, the prediction for future changes in hazard levels on account of climate change was broad-brush rather than detailed, as this level of detail requires such efforts as hydrologic/hydraulic modelling to consider how flood patterns may change under future scenarios. This certainly does not preclude such future elaboration of risk levels (i.e., detailed risk assessment) in the future on the part of government authorities and other stakeholders in the pilot cities. The findings of our analysis based on simpler approaches do provide useful guidance regarding the best use of funds for conducting such in-depth vulnerability and risk analyses (e.g., which hazards are likely to worsen, are there potential hotspots where hazards may get even worse, amongst others).

Urban, social, and economic adaptive capacity assessment
The availability of data meant that it was possible, within the timeframe, to conduct socio-economic assessment, and ascertain the exposure and sensitivity of urban residents to current and future flood and landslide hazards. An attempt was made to add to and to **thicken** the URA approach with more detailed consideration of the dynamics of both urban and economic growth, change, and development for cities. This involved studying the ways in which past and current social and economic processes have influenced spatial expansion patterns, which might have had and will likely continue to have an amplifying effect on flood and landslide

exposure. Adding this dimension makes assessment more dynamic (i.e., "adaptive") – accordingly, this assessment was re-titled (as above) to emphasise these urban and economic aspects.

As noted in the previous section, linking the potential climate projections to the way urban development is taking place is therefore essential in understanding the possible effects that climate variability and change could have. At different levels, the pilot cities have experienced both urban and economic transformations.

While Cusco and Castries strongly benefit from tourism, Santos is marked by the presence of a port and petrochemical activities, and Estelí and El Progreso are increasingly major industrial centres based on agro-processing. The impact that economic activity has on urbanisation is significant: economic growth attracts new populations through urban-rural migration. When combined with natural increase, this results in fast spatial expansion. However, spatial expansion was not always managed by appropriate planning instruments.

The cities often lacked the institutional structures and capabilities and the political will or backing to design and implement planning strategies that would give coherence to urban development. For instance, El Progreso benefited from industrial policies that fostered the emergence of manufacturing activities in the Sula Valley. Economic activity led to strong demographic pressures in the Sula Valley, giving rise to the emergence of a conurbation in the area (El Progreso Municipality 2012). Nevertheless, the various municipalities frequently lacked the suitable instruments to manage urban growth. This caused the emergence of dwellings in areas not destined for human settlement, notably in proximity to rivers and streams in the case of El Progreso, thus contributing to flood exposure.

The four other cities have gone through similar urban expansion processes which, in the absence of appropriate planning instruments, have at times led to habitation in areas not suited for it in the proximity of slopes and water streams, exacerbating landslide and flood exposure.

Poverty is embedded in the observed urban expansion processes and plays a major role in creating exposure to climate-related hazards. It is frequently the poorest populations that settle in the highest risk areas (Hardoy and Pandiella 2009; McGranahan, Balk, and Anderson 2007). The urban poor often lack the resources to access land markets in areas with lower or non-existent risk levels and they have to opt for affordable land in high-risk areas. This has been the case in Estelí, where low-income newcomers have located along the flanks of the Estelí River and its tributaries (Estelí Municipality 2008). Similarly, Cusco presents a complex inter-relationship between poverty, informality, and exposure to hazards. It is estimated that informal growth has accounted for around 80% of the recent growth in the city and a considerable group of newly-arrived migrants locate in areas not designated for habitation, for example, archaeological, ecological protection, forestry, agricultural, and hazard risk areas (Cusco Provincial Municipality 2006). Analysis further revealed that the areas marked by the highest exposure to climate-related hazards are located in districts with high levels of poverty, while the Wanchaq middle-class district is the only district in the city with negligible exposure to climate-related hazards.

Well aware of the inter-relationships between economic dynamics, urban expansion, and exposure to hazards, the municipalities, to varying degrees, have been proactive in establishing processes for developing planning instruments to guide urbanisation. For instance, with support from the World Bank, El Progreso has carried out land use assessments and attempted to incorporate risk into land use planning strategies. Santos has launched an ambitious municipal housing policy and relocation measures to remove all households from the high hazard risk areas. These are significant steps aiming to strengthen municipal urban planning capabilities and develop initiatives that integrate risk into land use considerations, with the overall objective of promoting sustainable urbanisation.

Institutional adaptive capacity assessment

The willingness of stakeholders to share their experience in planning, primarily for urban development and disaster risk rather than for climate change itself, permitted a full institutional assessment within the timeframe. This attempted to incorporate the dimension of institutional change, notably in the past decade, again to stress the element of dynamism that has (or may have) inhered to the institutions under study.

Although to different degrees, the history of all the pilot cities is marked by the recurrence of natural disasters, which have often caused severe impacts. These events have created a precedent by revealing disaster management and preparation shortfalls and triggering institutional change. This was very clear in El Progreso and Estelí where the devastation left by Hurricane Mitch in 1998 in Honduras and Nicaragua made evident the weaknesses of disaster risk management (DRM) systems in both countries, which then reformed their national systems in the decade that followed and established policy and legal frameworks for a comprehensive and multi-sector approach. Similarly, catalysed by fatalities experienced during landslide events in the 1970s and 1980s, the municipality of Santos developed comprehensive plans and procedures for risk reduction and management. The cities have thus learnt from their past experiences, and attempted to take action to tackle current challenges.

With the exception of Castries, where DRM decision-making is concentrated at the national level (the Castries City Council has both limited responsibilities and capabilities), a decentralisation process is implanted within the systems in place in the cities, and existing frameworks leave considerable room for action to local governments. Such frameworks establish the creation of local-level institutions responsible for the design, coordination, and implementation of DRM strategies. Such institutions are permanent, incorporate a wide range of stakeholders, and are often characterised by complex local-level governance structures.

As part of the National System of Disaster Risk Management (SINAGRED), within the Civil Defence in Peru, for example, regional and local governments are responsible for establishing working groups to develop and execute DRM strategies. Under the law, presidents of regional governments and local mayors are the apex authority in charge of supervising, leading, and executing DRM processes. The Cusco Provincial Municipality (MPC) actually leads the actions in the DRM system. Within the Provincial Municipality, the Committee of Civil Defence is the leading government structure in terms of disaster preparedness and relief. It is chaired by Cusco's Provincial Mayor. The Committee is a permanent government structure responsible for preserving the physical and material integrity of the population in the event of a disaster. It leads institutional coordination when emergencies and disasters occur.

If our analysis illustrated how cities are taking encouraging actions to address disaster management challenges, it also revealed shortcomings in institutional capacities. Of the most significance is the reactive approach that cities follow in disaster management. The focus tends to be on disaster mitigation, and risk management, in both policy and practice, is reactive and response-led. There are important efforts needed to enable a transition from disaster management to disaster risk reduction and climate change adaptation, particularly a shift to a proactive system tackling the roots that create exposure to climate-related hazards, and which permits establishing measures leading to long-term risk reduction. In this, Santos has progressed well. The city has established a progressive and adaptive approach to risk planning. Along with risk monitoring and strategic planning for flood and landslide hazards, the city's Civil Defence undertakes public education campaigns and direct community engagement in high risk areas. The organisation works in a pre-emptive manner by removing at risk populations both in advance of and during disaster events. The focus on risk prevention and reduction activities before the onset of a disaster, rather than only within the response phase, characterises risk planning in Santos.

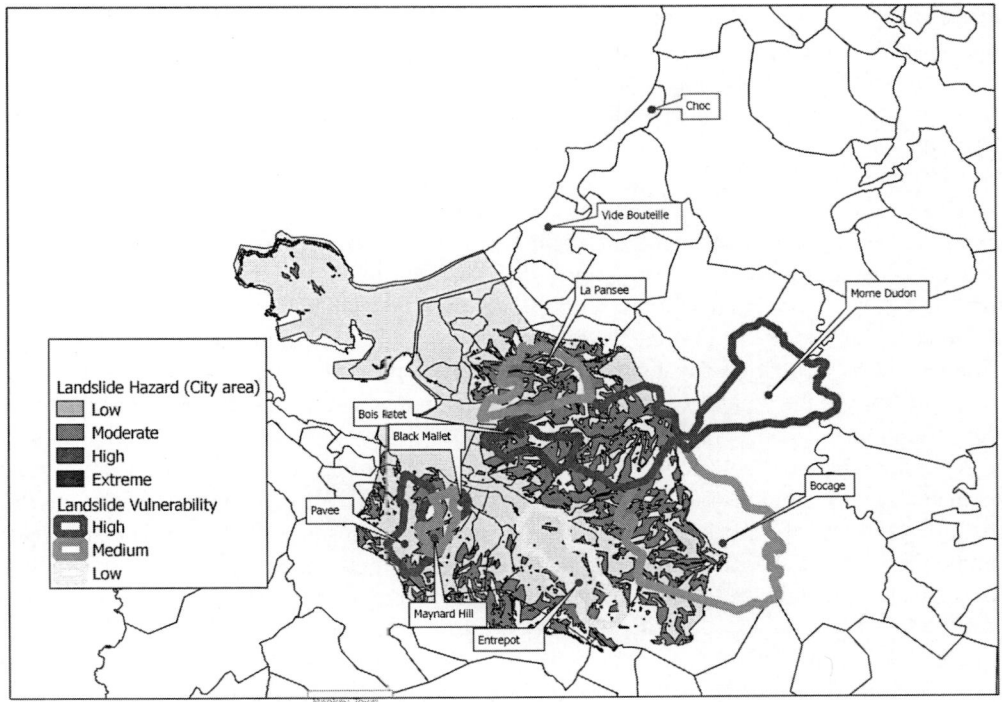

Figure 1.   Map of most vulnerable neighbourhoods and populations.

Climate-related vulnerability and risk assessment

To compensate for the limitations on risk assessment, a wider vulnerability assessment was developed than originally intended. This was effectively a summary of the findings of the three preceding assessments, and identified and mapped, to the degree possible, the most vulnerable city neighbourhoods and populations and exposed infrastructures to floods and landslides hazards, both currently and in the future (Figure 1). This was of necessity designed as an overview of vulnerability, rather than full assessment: as a vulnerability "screening" which could usefully be complemented by fuller and more detailed vulnerability analysis on the part of local stakeholders in the future.

A vulnerability analysis considers the exposure, sensitivity, and adaptive capacity of settlement to the hazard. The climate-related hazard assessment analysis identified the regions and settlements that were exposed to a landslide and/or flood hazard and then considered through the use of climate projections whether the exposure may increase or decrease by mid-century. Given that the hazard analysis did not include a more intensive modelling effort, the vulnerability analysis conducted was constrained to considering whether the existing hazard will worsen or reduce in areas already exposed to the hazard.

Because of this, this analysis could not provide a quantitative number describing the change in flooding or landslide exposure. However, it could certainly provide a high-level description of which settlements are exposed to the hazard and a qualitative description as to how climate change may impact future exposure.

Drawing then from the information available in the urban, social, and economic assessment, different metrics were used to as a proxy to suggest the proportion of settlements that may be able to better withstand landslides and to suffer less damage during a flood. For example, in Castries we used the percentage of households within each settlement fortified by concrete walls, while for Santos the sensitivity of each settlement was ranked based upon the population density and the percentage of households living in slums.

Adaptive capacity considers how a settlement that is exposed to and potentially harmed by the hazard may be able to cope or adapt. This may include considering what technological, economic, and social means are available to help the settlement to deal with the hazard. Again, the adaptive capacity within each settlement was based upon the available metrics in the cities. For Cusco, poverty levels and households with unsatisfied basic needs were used.

The vulnerability analysis then applied the rankings of sensitivity and adaptive capacity from low (i.e., least vulnerable) to high (i.e., most vulnerable) for each settlement that are located in flood and/or landslide-prone areas. The findings were discussed and "ground-truthed" during a workshop with primary stakeholders in all cities. This was an important step to ensure the rankings applied for sensitivity and adaptive capacity produced a credible vulnerability analysis.

A similar ranking of potential vulnerability based on sensitivity and adaptive capacity was not conducted for critical infrastructure facilities. The potential damage and associated costs to infrastructure – particularly infrastructure built to last many decades or more – is affected by many complex factors such as building materials and design, maintenance, age of the infrastructure, past damage, and the strength and dynamics of the particular event. Therefore, to analyse infrastructure vulnerability, a detailed database of these factors for each critical infrastructure is needed. Additionally, discussions with stakeholders can assist in understanding the level of difficulty in building protective resilience into the infrastructure.

Instead, a discussion was provided regarding the number of critical facilities located in landslide and/or flood-prone areas in each settlement as depicted in maps. The settlement's vulnerability to the event was also included in the maps to demonstrate the potential "social" severity if an event were to occur. For example, if a flood were to occur in a settlement that is highly vulnerable, there is a higher potential need for specific infrastructure – e.g., a hospital – to support the affected population compared to a settlement ranked at lower vulnerability. In addition, considering the criticality of the hospital to the local population is also important. These types of overlays can provide key talking points for decision makers when considering how to identify which infrastructure is critical to the local population and which thereby warrants further study.

Combined strategic climate adaptation investment and institutional strengthening plan

The preceding climate-related vulnerability and risk assessment provided the basis from which to identify and then prioritise a set of strategic climate adaptation investments and institutional strengthening interventions. A strategic, longer term view was proposed, coupled with action planning on a shorter time horizon in the short and medium term.

The plans drew on the conclusions and the feedback obtained during workshops held in all cities. There was enthusiastic participation by stakeholders in discussing initial assessment findings and in suggesting future strategy and concrete measures for adapting to current and future flood and landslide hazard risks. This interaction formed the basis for the plans. It should be emphasised that, by design, the plans have no particular institutional affiliation or official status – it, and the assessments and analyses upon which it is founded, now stands as a contribution offered to a debate that is already occurring and in fact is becoming more urgent in the pilot cities on climate change adaptation. Again, stakeholders in the cities will be able to adopt and elaborate the measures proposed as they see necessary.

The overreaching goal of the strategic plans was to propose an approach to increase resilience to floods and landslides in the cities. Specific planning themes for each city were determined to this end. These planning themes emphasised strengthening of urban planning and service delivery and municipal administrative capacity and resourcing. They were based on the priorities and main challenges identified in the assessments, which were later on confirmed or revised during the workshop. This process ensured that the suggested adaptation measures are suitable for the local context.

The planning themes led to the formulation of specific structural and non-structural measures which can be implemented to manage and reduce flooding and landslide vulnerability and risk. In broad terms, structural measures aim to reduce risk by controlling physical processes – such as the flow of water – both outside and within urban settlements. They are complementary to non-structural measures which aim at keeping people safe from flooding or landslides through better planning and management of – in this case, urban – development (Jha, Bloch, and Lamond 2012).

Although the planning themes varied, a number of challenges were shared by the cities. Three key examples were:

- Elaboration of land use planning systems, especially to understand the dynamics that trigger spatial expansion and to incorporate risk into land use planning strategies. In Cusco, for example, stakeholders acknowledged the need to improve planning instruments to identify areas that are at risk of flooding and landslides in order to design appropriate land use and expansion strategies to manage the settlement of new inhabitants. This could imply planting vegetation along the margins of rivers in order to enhance environmental protection as well as direct expansion away from risk areas. Additionally, stakeholders advanced the need to design potential relocation plans in areas where the risks cannot be mitigated.
- Improvement of mechanisms for data collection, storage, and dissemination for better climate monitoring, risk planning, and information sharing. This point was strongly present in Santos, as the possibility of establishing a City-Region Observatory was mentioned during the workshop. The observatory would work as a partnership between local universities, Santos Municipality, and other municipalities that make up the Baixada Santista Metropolitan Region, and would collect data, provide policy analysis and support, and undertake applied research.
- Consolidation of efforts to build capacity within, and resources for, city level government institutions in order to combine short-term disaster management with longer-term risk reduction and climate change adaptation. In Castries, the need to strengthen city-level institutions with responsibility for climate change adaptation and disaster risk reduction was recommended, as most of the decision-making in DRM is concentrated at the national level. This institutional change could enhance long-term risk reduction efforts by allowing a better implementation of climate change adaptation practices and policies.

In this plan-making process, the uncertainty associated with climate projections and its implications requires consideration. If uncertainty is not taken into account the risk is to prioritise investments and efforts that could lead to maladaptation. Maladaptation occurs when unsuitable investments are made for addressing the climate changes that actually do happen. In order to avoid it, flexibility should be incorporated into adaptation strategies by prioritising long-term adaptive capacity. In this, **robust** investments, as opposed to **optimal** ones, are to be prioritised. An optimal solution is only adapted for an expected future, but might be inappropriate if conditions change. Robust solutions on the other hand, might not be optimal, but they are appropriate no matter the conditions that are encountered in the future (Lempert et al. 2006).

The main challenge for planners and politicians is to implement a climate change adaptation process that considers the trade-offs between current development priorities and long-term climate risks. Uncertainty also needs to be acknowledged, as the timing and scale of local climate change impacts affects the types of measures to be adopted and prioritisation of investments and action. In the end, the ability and willingness of key political, economic, and technical actors to address climate change impacts will be of utmost importance.

Conclusions

The Climate Change Adaptation in Latin American and Caribbean Cities project allowed the application of the URA tool and the drawing of valuable lessons which may also be applicable to the many methods and tools available for climate change adaptation planning.

A first finding is that use of the URA tool demonstrated that detailed risk assessments require substantial technical capabilities and financial resources, as well as the commitment of time and the availability of significant amounts of data from actors and analysts alike. The division between primary, secondary, and tertiary levels of assessment, described above, is helpful in this regard and could even be elaborated further so that stakeholders are aware at exactly at what level of detail the analysis and assessment will take place. Here, crucially, the identification and hence **existence** of data is not however equal to **the availability** of data: for a variety of institutional, political, and even personal factors, data might be unavailable, difficult to get, or in a form that is difficult to analyse. It would be ideal to have access to all existing data/information, as identified, but in reality this is not often possible.

Secondly, **urban** considerations somehow disappear in many urban risk assessment tools, including the URA tool, and as they are carried out they become what can be seen as **standard** assessments. It is necessary to be alert to this and to bring back and highlight urban development considerations and trends, linking the dynamics of urban growth and development and spatial expansion to socio-economic, institutional, and hazard analysis (a key aim should be to assist cities to avoid risk areas as they expand).

Thirdly, the URA tool is largely silent on political interests and the dynamics of political (and economic) contestation. Without overburdening them, the conducting of risk assessments would benefit from more formal consideration of the political economy context within which CCA operates in any given situation (trade-offs in public investment decisions, intra-governmental relations and local governance, and other formal and informal political and institutional structures and processes, to name but few issues). To put it concisely, just as urban development dynamics must not be overlooked, neither should urban politics.

Finally, adaptation planning needs to be able to learn from planning processes over time and to adjust methodologies. There are many available methods and tools for climate change adaptation. Many of these do not automatically work in the rapidly urbanising cities of developing countries. The URA does. Our experience suggests that catering for a process by which local knowledge and more formal technical knowledge and assessment techniques are integrated is essential. A successful process requires time and the commitment of real resources from a city. That said, there must always be a focal institution, and if you work with cities you need to work with local/municipal governments – and more specifically, with the particular departments or agencies responsible for urban development, disaster risk, climate change, and environment, and coordinating between them.

In the short and medium term, our experience demonstrated that cities, particularly those with urgent development needs, need to prioritise actions and investments that deal with current hazards risk and offer development co-benefits. In an incremental process, city governments, as they develop capacity and more data becomes available, can focus and deal with future risk and prioritise actions and investments that seek to address the adverse impacts of hazards and related disasters increasingly associated with climate change. That said, the translation of findings to implementable policy, strategy and action planning – as the next step and key outcome of such processes – is an area in which there is scope for more "how to" guidance for cities.

Acknowledgements

The paper is a product of the Climate Change Adaptation Planning in Latin America and Caribbean Cities initiative led by the World Bank's regional Urban and Disaster Risk Management Unit for Latin America and the Caribbean (LAC) (LCSDU). The findings, interpretations, and conclusions expressed in this paper are entirely those of the authors. They do not necessarily represent the views of the World Bank and its affiliated organisations, or those of the Executive Directors of the World Bank or the governments they represent.

References

Bloch, R., N. Papachristodoulou, and D. Brown. 2013. "Suburbs at Risk." In *Suburban Constellations. Governance, Land and Infrastructure in the 21st Century*, edited by R. Keil, 95–101. Berlin: Jovis.

Cusco Provincial Municipality. 2006. "Cusco Province Urban Development Plan 2006–2011."

El Progreso Municipality. 2012. "Municipal Development Plan with Territorial Planning Focus. Annex 6 Special Report on Risk Management."

Estelí Municipality. 2008. "Estelí Municipal Diagnosis."

Hardoy, J., and G. Pandiella. 2009. "Urban Poverty and Vulnerability to Climate Change in Latin America." *Environment and Urbanization* 21 (1): 203–24.

Jha, A., R. Bloch, and J. Lamond. 2012. *Cities and Flooding: A Guide to Integrated Urban Flood Risk Management for the 21st Century*. Washington, DC: The World Bank.

Jha, A., J. Lamond, R. Bloch, N. Bhattacharya, A. Lopez, N. Papachristodoulou, A. Bird, D. Proverbs, J. Davies, and R. Barker. 2011. "Five Feet High and Rising: Cities and Flooding in the 21st Century." World Bank Policy Research Working Paper 5648. Washington DC: The World Bank.

Lempert, R., D. G. Groves, S. W. Popper, and S. C. Bankes. 2006. "A General, Analytic Method for Generating Robust Strategies and Narrative Scenarios." *Management Science* 52 (4): 515–528.

McGranahan, G., D. Balk, and B. Anderson. 2007. "The Rising Tide: Assessing the Risks of Climate Change and Human Settlements in Low Elevation Coastal Zones." *Environment and Urbanization* 19 (1): 17–37.

Seto, K. C., B. Güneralp, and L. Hutyra. 2012. "Global Forecasts of Urban Expansion to 2030 and Direct Impacts on Biodiversity and Carbon Pools." *Proceedings of the National Academy of Sciences of the United States of America*.

Seto, K. C., R. Sánchez-Rodríguez, and M. Fragkias. 2010. "The New Geography of Contemporary Urbanization and the Environment." *Annual Review of Environment and Resources* 35: 167–94.

UN. 2012. *World Urbanization Prospects: The 2011 Revision*. New York: United Nations Department of Economic and Social Affairs.

UN-HABITAT. 2012. *The State of Latin American and Caribbean Cities 2012: Towards a New Urban Transition*. Nairobi: United Nations Human Settlements Programme.

UNISDR. 2011. "Global Assessment Report on Disaster Risk Reduction: Revealing Risk, Redefining Development." Geneva: United Nations International Strategy for Disaster Reduction.

World Bank. 2011. *Urban Risk Assessment: An Approach for Understanding Disaster & Climate Risk in Cities*. Washington, DC: The World Bank.

Institutionalising mechanisms for building urban climate resilience: experiences from India

Anup Karanth and Diane Archer

This paper examines how mainstreaming of urban climate change resilience – a crucial consideration in an increasingly urbanised world – is occurring at both the city and national scale, using the case of an internationally-funded resilience-building initiative in India. Surat city's newly-established Climate Change Trust illustrates the importance of an institutionalised mechanism for coordinating and sustaining climate initiatives. Concurrently, climate resilience is being mainstreamed into the national urban development agenda, through a network of Indian institutions. These two nascent mechanisms offer avenues for local city-level experiences to inform national directives, driving and sustaining the urban climate adaptation agenda across India.

Cet article examine comment l'intégration de la résilience au changement climatique en milieu urbain – une considération cruciale dans un monde de plus en plus urbanisé – a lieu à l'échelle des villes et au niveau national, en se servant du cas d'une initiative de renforcement de la résilience financée par des sources internationales menée en Inde. Le Climate Change Trust, récemment établi dans la ville de Surat, illustre l'importance d'un mécanisme institutionnalisé pour coordonner et soutenir les initiatives liées au changement climatique. Dans le même temps, la résilience au changement climatique est intégrée dans l'ordre du jour de développement urbain national, à travers un réseau d'institutions indiennes. Ces deux mécanismes naissants offrent des voies permettant aux expériences locales au niveau des villes d'éclairer des directives nationales, impulsant et maintenant l'ordre du jour relatif à l'adaptation au changement climatique en milieu urbain aux quatre coins de l'Inde.

El presente artículo examina de qué manera la resiliencia ante el cambio climático es incorporada a nivel urbano y nacional –acción crucial en un mundo cada vez más urbanizado. Para tal efecto, se analizó una iniciativa financiada con fondos obtenidos a nivel internacional, dirigida a fortalecer la resiliencia en India. El recientemente establecido Fideicomiso para el Cambio Climático en la ciudad de Surat, da cuenta de la importancia de contar con un mecanismo institucionalizado para la coordinación y la sostenibilidad de las iniciativas en torno al clima. Paralelamente, a través de una red de instituciones indias, la resiliencia ante los cambios climáticos se está incorporando a la agenda de desarrollo urbano en todo el país. En India, estos dos mecanismos emergentes permiten que las vivencias a nivel urbano influyan en las directrices nacionales orientadas a impulsar y sostener esfuerzos cuyo objetivo apunta a instrumentar adaptaciones a los cambios climáticos a nivel urbano.

CLIMATE CHANGE ADAPTATION AND DEVELOPMENT

Introduction

The impact of climate change poses considerable risk to Indian cities. It is expected that impacts will include: a general increase in temperatures by 2–4°C, an increase of 7–20% in annual precipitation, with increased intensity, alongside increases in riverine flooding, cyclones, storm surges, and sea-level rise (SLR) (Revi 2008). A study on Low Elevation Coastal Zones estimates that India sees about 3% of its national area at risk of SLR (McGranahan et al. 2007). All of these impacts come with associated environmental health risks. Climate change impacts in Indian cities need to be considered in the context of significant demographic, rural to urban, and environmental transitions (Revi 2008). As per the Census of India Report 2011, the urban population of India has increased from 285 million in 2001 to 377.1 million in 2011. In 2011, the urban population was 31.15% of the country's total, as compared to 27.8% in 2001. It is expected that 500 million people will move from rural to urban areas by 2060 (Revi 2008). Given that urban areas concentrate disaster risk, this continuing urban growth poses significant challenges for effective and adaptive urban management in Indian cities.

The populations most vulnerable to climate change induced hazards are often those in informal settlements, who live in insecure housing, with inadequate or non-existent provision of basic services, and frequently along the marginal and most at-risk areas of a city, such as on riverbanks. At the same time, vulnerability is not solely dependent on hazards but also wider socio-economic risk (Satterthwaite et al. 2007). In order to reduce vulnerability, a shift in public policy from greenhouse gas mitigation to incorporate adaptation is required, with action and the necessary structures and institutions in place on multiple levels, from the national to the state, city, and neighbourhood levels (Revi 2008). Implementing initiatives to adapt to climate change in cities allows scope to develop and test integrated and multi-stakeholder approaches which can be scaled up and integrated with existing national programmes for both disaster risk reduction and urban development.

In light of the need to take action to respond to climate change, an initiative called the Asian Cities Climate Change Resilience Network (ACCCRN), funded by the Rockefeller Foundation, has sought to make urban climate resilience an integrated and sustained component of city development activities. The initiative began in 10 core cities across four Asian countries (India, Vietnam, Indonesia, and Thailand), and the approach is now being up-scaled to a further 30–40 cities, including in two more countries, the Philippines and Bangladesh. In India, ACCCRN began in 2008 with activities to develop city resilience strategies in three cities: Surat, Indore, and Gorakhpur, and is now being expanded to a further 24 cities, alongside activities to institutionalise and mainstream considerations of urban climate change resilience (UCCR) both at a city and national scale. This paper will examine how this transition into mainstreaming UCCR is happening by looking at the journey of ACCCRN in India and how it is scaling up beyond the city level to networking and coordination at the national level, using the experience of Surat and its newly-established Climate Change Trust as a city case study.

Governance for urban climate change resilience in India

As the impact of climate change-induced hazards in India is very much linked to the existing structural vulnerability of a large part of the urban population, adaptation to climate change cannot be achieved without addressing the institutional weaknesses in managing urbanisation and ensuring service delivery, alongside the necessary planning and regulatory frameworks (Revi 2008). India's governance structure means that there are entry points for climate-change related action at the national, state, and city levels, as illustrated in Figure 1.

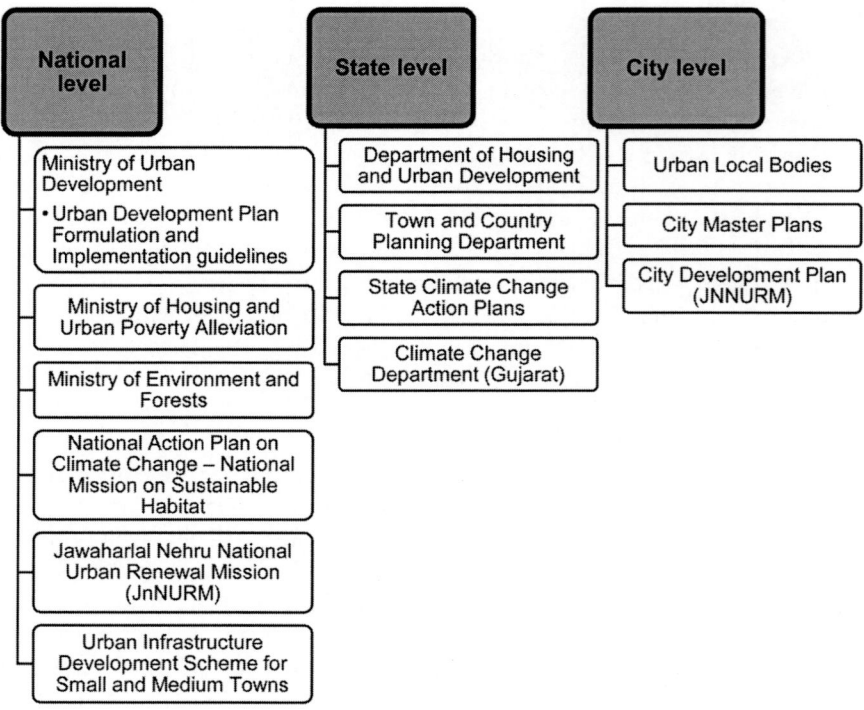

Figure 1.   Stakeholders and entry points for climate change resilience planning.
Note: Modified from TERI 2011, 5.

India's National Action Plan on Climate Change (NAPCC) issued in June 2008 offers opportunities for integrating urban resilience planning, through the eight National Missions, such as the National Mission on Sustainable Habitats (NMSH) and the National Water Mission (TERI 2011). There is also the opportunity to make use of existing national urban renewal programmes, such as the Jawaharlal Nehru National Urban Renewal Mission (JNNURM), which focus on basic services for the urban poor, and urban infrastructure and governance, but originally included no components on climate adaptation or risk mitigation (Revi 2008). As the state level also influences both urban development and climate change adaptation planning, it is an important actor in setting the necessary policy framework for city-level action. City development plans are also an avenue for developing an action plan to address inequalities in service provision. However, for this to be effectively implemented, there is a need for adequate capacity building at the city level with regards to understanding climate risks and how they can be mitigated, and effectively coordinated across the multiple stakeholders and agencies which need to be involved.

Urban climate governance literature highlights the need for multiple levels of actors and institutions to be engaged (Corfee-Morlot et al. 2011; Brown, Dayal, and Rumbaitis del Rio 2012; Leck and Simon 2012). While many local level climate change initiatives are undertaken 'autonomously', multilevel or multiscale governance is necessary to address mismatched priorities across institutional bodies (Leck and Simon 2012). Multilevel governance should also extend into opportunities for bottom-up, community-based adaptation activities to be supported and linked into wider city-level adaptation plans or initiatives. At the same time, the appropriate institutional coordination mechanisms and capacity support are required to support climate adaptation planning activities locally (Brown, Dayal, and Rumbaitis del Rio 2012). Moving from the local to the national, Fisher (2012, 125) sees a growing role for loose networks of non-state and subnational actors to govern aspects of climate change through *"norm diffusion"* to those working on development projects on the ground.

Given the many competing pressures for time and resources which city actors face, drivers for taking action on climate change adaptation also need to be considered. Bulkeley and Castán Broto (2012) see urban climate change initiatives as forms of governance experimentation, which may be shaped by international, national, and local policy drivers. Carmin, Anguelovski, and Roberts (2012, 20) argue that with regard to climate adaptation, exogenous forces such as external funding, national funding, and local champions need to be supplemented by endogenous forces such as municipal commitment and local civil society action and pressure – and that, in an emerging field such as adaptation with few models to be emulated, exogenous forces are more likely to play a role at later stages, while local leaders and other endogenous factors will be the initial drivers. Similarly, Kernaghan and Da Silva (2014) identify sustaining factors at the city level and beyond the city level, drawing on ACCCRN processes across Asia. These factors fall under four key themes of knowledge, finance, policies and plans, and stakeholders, which contribute to creating an enabling environment for developing resilience.

As the case of Surat below illustrates, the city's strong industrial base has made the Southern Gujarat Chamber of Commerce and Industry a champion of city-level action for UCCR. Hence, action on climate resilience is in this case driven by a mix of endogenous and exogenous forces, with prior experience of disasters and external funding giving the city the push it needed to comprehensively consider and strategise on resilience, while endogenous forces are now sustaining action. At the same time, in the absence of international best practice, capacity, and resources, innovation can emerge as a response to the need for climate action (Anguelovski and Carmin 2011). As a result, Surat is one of India's early adopters, and can thus present a model for other cities to emulate, whilst recognising the particular characteristics which have enabled the city to advance the resilience agenda.

Surat – balancing investments and risks

Surat is one of India's most progressive cities. Located in a coastal area of Gujarat state, it has a population estimated at 4.5 million in 2011 (Surat City Resilience Strategy 2011), making it the ninth largest city in India, and is a major commercial and industrial centre, its main industries being rough diamond cutting and polishing, textiles, machinery, and chemical. The city is seeing a major investment drive, with massive infrastructure projects, land reclamation, a SEZ, and tax incentives; however, this investment needs to be balanced with a consideration of risk, particularly climate change associated risks. The city's industrial base also draws a large migrant population of workers. In 2006, the city was hit by a massive flood which inundated 75% of the city's area, with a huge associated economic cost, and the loss of 150 lives (though unofficial estimates put the death toll at up to 500 persons). Floods in the city have two main causes: Ukai floods, which come from release of water from the upstream Ukai dam 100 km away and can be rapid and severe (especially when coinciding with high tides); and *khadi* floods, which are more frequent but less devastating, caused by the overflow of two streams running through the city. As many of the city's poorer groups live along these creeks, they are particularly at risk from *khadi* flooding. The Surat Municipal Corporation has undertaken efforts to shift the people from high risk prone areas to a safe location, through state-sponsored social housing schemes such as Valmiki Ambedkar Awas Yojana (VAMBAY), the JNNURM, and more recently, the Rajiv Awas Yogna (RAY) scheme.

A hazard, risk, and vulnerability analysis carried out in 2005 showed that Surat is a major hotspot for hydro-meteorological hazard risks (Gujarat State 2006). It is estimated that around

Figure 2. Expansion of Surat city limits (1963–2006).
Source: TARU Leading Edge.

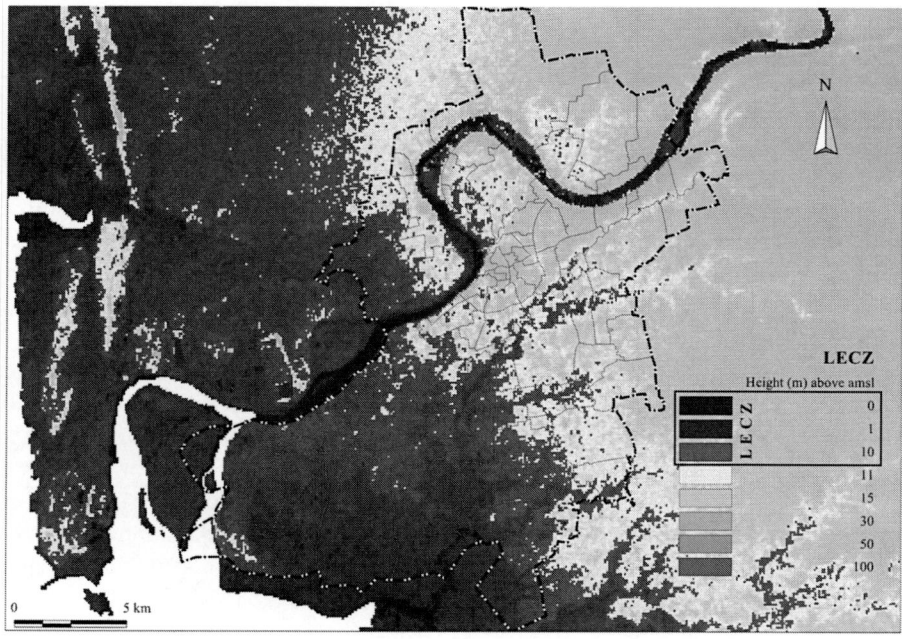

Figure 3. Map of low elevation coastal zones of Surat.
Source: TARU Leading Edge.

90% of Surat's geographical area is affected by some type of climate hazard, which includes flooding, coastal storms and cyclones, and inundation associated with high tides and SLR (Bhat et al. 2013). Much of the city and its surrounds are less than 10 m above mean sea level, and a one metre SLR would swamp 40% of the land area (Figures 2 and 3). A major new industrial investment area, Hazira, on a flood plain on the periphery of Surat, has already suffered extensively from the massive 2006 floods and is at risk from further flooding, cyclonic storms, and SLR.

In 1994, Surat faced an epidemic of the plague. This epidemic served as a tipping point for the city to overhaul its health systems and ensure better coordination across agencies and actors. Similarly, the 2006 floods can be seen as a tipping point, impacting physical assets and production, causing losses totalling Rs160 billion (around US$3.5 billion), of which three-fifths were direct losses and the rest from loss of production, while losses to infrastructure including the dam, flood embankments, electricity, and telephone lines totalled Rs25 billion (US$544 million) (Bhat et al. 2013). Together, these two catastrophic events changed the city's dynamics to create inherent built-in resilience, which can be leveraged to further develop the city's resilience to climate change.

Any discussion and consideration of climate change risks taking place within Surat, and how they might be mitigated through adaptive actions, should therefore balance the actions of major business actors, who are driving the city's growth, alongside the need to reduce further exposure and risk, and protect vulnerable populations, who may be key workers in the city's industry and who are often most badly hit by the floods. As Surat's flood risks derive in part from the Ukai dam, which is in the catchment area of three states, actions need to be integrated and involve actors beyond the city's boundaries. Bridging risk and investment priorities is also an important consideration, which should be led by a body or actor with the capacity, and ability, to send clear messages to all stakeholders in the city's urban and economic development.

Building urban climate change resilience in Surat

The ACCCRN initiative in Surat began in 2009 by the establishment of the City Advisory Committee (CAC). The CAC mechanism brought together key stakeholders in developing and managing the city's resilience-building activities, including the Surat Municipal Corporation (SMC); the city Commissioner; the South Gujarat Chamber of Commerce; and industry, educational institutions, and other actors. The CAC received good feedback as a simple mechanism for getting the necessary engagement on climate change issues at the city level. This was followed by conducting detailed vulnerability assessment processes across all socio-economic groups (see Bhat et al. 2013), which considered both adaptive capacity (with regard to education, income stability, social capacity, and access to loans and insurance), and vulnerability (in relation to infrastructure, particularly drainage and sewerage, water, and floods). The vulnerability assessment process fed into the development of the Sector Studies on urban environment, public health, energy security, water security, and flood risk management. The Surat City Resilience Strategy (2011) was informed by a range of studies and results of several workshops and consultations. Scenarios were constructed of possible futures under the changing climate. The participatory process built up a sense of engagement with a wide range of stakeholders and the disciplined process of thinking ahead engaged them to together identify key sets of actions. The strategy document, which is owned by the CAC and the Surat Municipal Corporation, outlines short, medium, and long-term priority actions to address climate change risks and minimise impacts. The strategy is structured around three principles: build upon current and planned initiatives, demonstrate resilience-building projects in order to leverage further action, and build synergy with state and national-level institutions (TERI 2011).

From the City Resilience Strategy, a range of resilience interventions were prioritised and implemented through ACCCRN pilot programme support. These include:

- the Urban Services Monitoring System (UrSMS), for improving monitoring and grievance redress for health, water supply, sewerage and solid waste services. This system can be upgraded for use during emergencies such as floods;
- the development of an end-to-end early warning system for Ukai and local floods, based on an integrated modelling system. At the same time, the capacity of civil society, businesses, and city institutions to prepare for and manage flood emergencies;
- the establishment of Urban Health and Climate Resilience Centre (UHCRC) to carry out disease surveillance, with a particular emphasis on climate-related illnesses and poor and vulnerable populations in the city; and
- promotion of cool roofs and passive ventilation concepts for indoor temperature comfort.

During the implementation of the second phase of ACCCRN activities, two groups were formed at the city level: the Climate Change Watch Group and the Awareness Generation Group. Both the groups undertook a range of initiatives in the city and promoted climate change as an important issue. While these groups have now faded out, they achieved their role of building awareness of the need to consider and tackle climate change risks in the city. Individuals who volunteered their time earlier are now involved in mainstreaming climate change issues in various development activities of the city, particularly by raising awareness and literacy around climate change, with considerable influence on decision-making at city level.

In 2010, a State consultation meeting was held to understand the dynamics of managing floods in Surat. Key stakeholders, including the CAC and managers, deliberated and realised that there were close to 12–15 actors and institutions with flood-management responsibility in the city. The consultation meeting gave an opportunity for these stakeholders to discuss approaches to effective flood management. It became apparent during discussions that a mechanism for such multi-stakeholder deliberations should be established to facilitate regular meetings of these actors. This would be key in supporting the integrated design of the flood management system, given the trans-boundary nature of the river system and the various responsibilities of the many organisations involved. A collaborative mechanism would also be able to direct agencies in data collection and collaborate on the various other aspects of the end-to-end early warning system. This was the seed for the establishment of the Surat Climate Change Trust (SCCT). While a number of possible structures were mooted for the shape of this mechanism, a registered Trust was chosen as the form best suited to managing a project that would span a range of institutions and actors beyond the city boundaries.

The SCCT initially therefore began as a vehicle for implementing the end-to-end early warning system. The SCCT has taken the lead to implement other ongoing ACCCRN projects and discussions are being held to continue to undertake a range of interventions as outlined in the City Resilience Strategy. TARU Leading Edge, a private company which is the Country Coordinator of the ACCCRN processes in India, played a vital role as a facilitator and catalyst in the formation of the SCCT, and in driving ACCCRN activities in Surat and Indore.

The governance of the SCCT

The Surat Climate Change Trust (SCCT) is a city level public trust registered under the Bombay Public Trust Act 1950 (Registration No. E-7266/Surat) with its office at the City Engineer's Office, Surat Municipal Corporation. While this process of registration caused significant delay in the implementation in the flood early warning project, it was strongly felt that the real

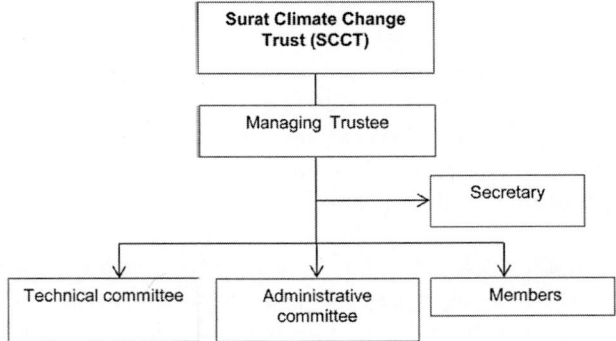

Figure 4.   Surat Climate Change Trust operating structure.

sustainability of the project required this high end mechanism in place beforehand, with all the necessary regulatory and financial provisions met. The Trust's management and operating structure is illustrated in Figure 4.

The Trust's board is made up of 13 members, of whom six are from the Surat Municipal Corporation, including the city's Commissioner, Deputy of Policy, City Engineer, Deputy Commissioner, and Corporators. The other board members include the President and Vice-President of the South Gujarat Chamber of Commerce, which was instrumental in driving ACCCRN processes forward in Surat, and representatives from academic institutions and state level stakeholders – the Narmada Water Resources, Water Supply and Kalpsar Department, the Gujarat State Disaster Management Authority, and the Urban Social Health Advocacy and Alliance (USHAA). The Surat Municipal Corporation (SMC) promoted the process of formation of the Trust, and provides a firm footing or anchor for the SCCT. The make-up of the board facilitates linkages to key actors and institutions in the city, as well as beyond its borders – a particularly important consideration for water management.

The technical committee plays an important role in ensuring due diligence in processes such as tendering for bids, which are then considered by the members. The structure of the SCCT means that it has the capacity to manage project implementation. Importantly, it also has the ability to receive funds direct from donors or other actors. This is a crucial consideration in a context of climate financing mechanisms which are predominantly designed to pass through national governments, reducing the options of local governments in securing adequate financing. Thus, while still in its early days, the SCCT will play an important role as an institution enabling the city to receive funds for climate change adaptation activities from various sources. The SMC has also earmarked funds for the SCCT, thus giving the body an early ability to stand on its own two feet.

Role of the trust

As outlined below, the objectives of the Trust relate to advocating for and implementing projects and activities in support of urban climate resilience. This includes activities relating to the mitigation of greenhouse gas emissions, as industry is an important part of the city's economy and growth. The SCCT will also gradually build its capacity to communicate climate change issues effectively, and has just launched a website which provides resources on climate change mitigation and adaptation (www.scctrust.in).

The objectives and visions of the Trust are:

- to **engage** in policy advocacy regarding urbanisation and climate change;
- to **facilitate capacity building** of the City of Surat to face challenges of urbanisation and climate change and to **facilitate development** of a roadmap to face these challenges;
- to **achieve** stabilisation of greenhouse gas concentrations in the atmosphere at a level that would prevent dangerous anthropogenic interference with urbanisation and the climate system;
- to **support** and to **undertake measures and projects**, which minimise urbanisation and climate change damage by reducing its risks or its adverse effects;
- to **support** and to **undertake interventions** that increase resilience of vulnerable sectors and communities in the society to the adverse impacts of urbanisation and climate change;
- to **spread awareness** about techniques, technology, and practices related to urbanisation and climate change and to undertake and to facilitate urbanisation and climate change awareness and training for different sections of citizens;
- to **impart best quality education** including moral, spiritual, physical, commercial, industrial, scientific, medical, and technical education, about urbanisation and climate change;
- to **conduct** various public welfare activities;
- to **conduct and promote** various activities for social development and improvement of quality of life of citizens; and
- to **give aid** to any other charitable institutions having similar objectives.

Increasingly, it was felt that the SCCT could be an effective mechanism for carrying forward the urban climate change resilience agenda, by promoting resilient urban development in the Surat Metropolitan Region, as the larger area that Surat city will grow to occupy over the next decades. Moving beyond the end-to-end early warning system, a second ACCCRN intervention, the Surat Urban Health and Climate Resilience Centre (UHCRC), was launched in March 2013. The implementation of the UHCRC is managed through the SCCT, and the Centre will function not as a stand-alone institute but as part of existing entities, hence ensuring its sustainability.

These two ACCCRN-funded interventions, as fairly soft projects, will help develop the capacity of the SCCT in managing interventions to build urban climate resilience, while looking ahead to attract funding for further projects from other sources. They build upon the pilot projects as outlined above. As the SCCT includes representatives from many institutions, it can help to bridge the gaps between different organisations involved in the development of Surat. Because the SCCT includes representatives from local government and local actors, it can also obtain a more rapid response from key government departments at State or National level, such as the Indian Meteorological Department and the Central Water Commission. These are vital institutions for the implementation of the flood early warning project, thus addressing one of the key challenges of coordinating complex multi-stakeholder and multi-scalar initiatives. In the other direction, the existing City Disaster Management Plan will also undergo significant modification, as there is an increased emphasis on connectivity arising from the experience with the SCCT. As a result, the linkages between science, institutions, and society will be strengthened at the city scale.

As the Trust also includes two representatives from the Southern Gujarat Chamber of Commerce and Industry (SGCCI), the city's activities relating to building climate change resilience are shared with a range of actors beyond the public sector, while also linking the public and private sectors, facilitating the city's progressive actions. As the business community is strongly engaged in the city's investments in high-value infrastructure, and it requires continuity in its activities, there is an incentive for it to be engaged in the city's efforts to reduce disruption from risks such as flooding and sea level rise. The Trust, as an institutionalised mechanism, offers the opportunity to ensure adaptation activities can be sustained beyond electoral cycles, which can be a barrier to sustained climate action at city scale (Brown, Dayal, and Rumbaitis del Rio 2012).

The potential for replication

It is worth examining more closely the factors which have enabled Surat's success in establishing a Trust which should be a driving force in sustaining UCCR action in the city. The potential for replicating a SCCT-type model in other cities is contingent upon a number of factors, of which the local institutional, economic, and demographic contexts are important considerations. Firstly, Surat's strong industrial and business sectors mean that there are incentives for these sectors to be represented within the Trust through the Chamber of Commerce, securing a broader base beyond public sector representatives. Additionally, flood management considerations mean stakeholders beyond city boundaries are also involved. This broad, multi-stakeholder representation has implications for the sustainability of the Trust beyond local electoral cycles, and could be applied to other cities facing trans-boundary climate risks such as flooding or drought.

The long timeframe for the institutionalisation of the SCCT, of over three years from initial discussions to legal registration, may, in other circumstances, have meant the initiative lost its momentum – however, non-political actors can drive forward their agenda where government actors might otherwise face changing priorities. Additionally, the role of exogenous forces, particularly the technical support and funding for projects through ACCCRN, spurred on the process of institutionalisation. In other circumstances, the momentum might be sustained with pressure from civil society groups or local communities experiencing the climate impacts themselves, if they are sufficiently organised and able to work effectively with the local government. If the SCCT demonstrates itself to be successful in securing funding from donors, private sector or government sources, this may encourage other cities to emulate the Trust in order to enable more independence in addressing local climate priorities.

The city's large industrial sector facilitates a strong municipal tax base, enabling a degree of financial autonomy from which other cities may not benefit, and which also makes the Chamber of Commerce a key stakeholder in the city's growth and development, given the large-scale economic losses faced by businesses resulting from previous disasters. As such, the Chamber of Commerce can be considered a key champion driving forward the process, especially as the projects in progress, for health and flood management, are relevant across the city, rather than only specific areas or communities. In other cities, if those most affected by climate change are urban poor or marginalised groups, their ability to drive forward a resilience agenda at city scale may be limited unless linkages to other sectors of importance to the city are highlighted. Having a champion on-side, whether an elected representative or from a civil society organisation, can help to raise the issue. For example, in Gorakhpur, another ACCCRN city, local communities have played an important role in setting and addressing local priorities, with support from Gorakhpur Environmental Action Group (GEAG), the local ACCCRN partner. Communities in a particularly climate-vulnerable ward established committees at the neighbourhood, ward, and city levels around thematic issues such as health, water, and sanitation. These committees feed into a City Steering Committee and other city-level stakeholders, thus reshaping processes of urban governance, and are composed of a broad range of representatives ranging from academia to civil society (Mani and Wajih 2014).

Scaling up UCCR in India from city to national level

The case of Surat has illustrated the importance of having an established and institutionalised mechanism for coordinating UCCR activities at the city level, but also highlighted the unique factors which framed the city's approach to building resilience. As ACCCRN enters its third and final phase in India, the focus is now turning to building up a mechanism for scaling up and mainstreaming considerations of urban climate change resilience into national activities. This is being facilitated through a Consultation Committee of 11 institutions involved in urban development and climate change adaptation activities. These 11 organisations, working as a network, will promote resilient urban development through a series of national, state, and city-level consultations and events, building on the ACCCRN approach applied at the city level. The reach of ACCCRN in India now extends to 24 cities, with 12 partner organisations playing a role in this expansion. As per Kernaghan and Da Silva (2014), networks external to a city can play a key role in knowledge dissemination and sharing experiences more broadly, in support of action at city scale.

As the Government of India seeks to promote sustainable urban development through programmes such as the JNNURM, the RAY programme for a slum-free India, and the National Sustainable Habitat Mission, there is an opportunity to ensure that these programmes and initiatives integrate climate change considerations to ensure sustainable initiatives. According to Revi (2008, 211), a "*chasm exists between the official urban 'city building' development agenda and vulnerability reduction for those most at risk*", so these national programmes are an appropriate avenue for integrate climate change adaptation concerns. The Revised City Development Plan Toolkit (of JNNURM) now includes a mention of climate change under clause 6.4.13. (Ministry of Urban Development 2013).

The national Consultation Committee sees a need to shift the focus from looking at cities as purely driven by commercial needs, to places with the capacity to optimise their internal resources and make the city development process more vibrant whilst also inherently resilient. In order for cities to be supported in adaptation planning, it is necessary to also engage at the state and national level to develop policy which will frame and direct action at city levels (Sharma and Tomar 2010).

In order to achieve this goal, the Consultation Committee functions as a forum for addressing the key issues, challenges, and processes related to promoting resilient urban development, while strengthening and complementing each other's efforts to achieve urban climate change resilience. The Committee will also serve as an advocate for urban climate change resilience, and seek to strengthen existing national urban development programmes in this regard. This will require a training and capacity building component for city stakeholders in order to develop their understanding of climate change impacts and how they can be addressed. Additionally, a network for cross-learning between institutions and their partners will be facilitated, with an emphasis on policy lessons relevant to urban development. The existing Peer Experience and Reflective Learning (PEARL) network, housed within NIUA, is a possible avenue for this networking. As many of the members of the Committee have experience in implementing UCCR-related measures at the city level, either through ACCCRN or other initiatives, their experiences can be shared to develop methodologies for resilient city development. Another component will be further research and documentation around issues pertaining to UCCR in order to better inform actions. This represents a route for engaging in the policy process through strategic communication, and problem and solution coupling (Fisher 2012).

The establishment of the Consultation Committee has been facilitated by the ACCCRN programme. As ACCCRN is an independent initiative, it has the advantage of functioning as a uniting mechanism bringing together various partners. A number of the Consultation Committee members (GEAG, ICLEI, and TERI) are also supporting city-level initiatives in Indian cities with ACCCRN support, testing different approaches and tools, and these experiences will inform policy lessons and capacity building. These organisations also regularly engage with city-level institutions and actors and can facilitate multilevel or multi-scalar linkages across local, state, and national organisations.

Lessons learnt in institutionalising and scaling up UCCR initiatives in India

With climate change and urbanisation coming head to head, the need to consider measures for building urban climate change resilience alongside urban development is increasingly becoming apparent. For India, partnerships across disaster management and other institutions, including in environment, health, and planning, will be required to facilitate urban adaptation planning (Sharma and Tomar 2010). Action needs to be taken at multiple levels, and capacity to implement and manage adaptation programmes at all levels therefore needs to be taken into consideration when developing measures. The SCCT has emerged as a mechanism which can forge partnerships across multiple institutions, including from both the private and public sector. However, it must be recognised that while the SCCT can be an effective mechanism for linking the various decision-making bodies at city scale, the role of civil society in urban governance should not be neglected. Mechanisms for the involvement of the wider urban population in urban climate adaptation planning processes should also be considered, as in the case of Gorakhpur, where local communities have been actively driving adaptation initiatives in their neighbourhoods.

The process applied in Surat in developing the SCCT has received recognition from the national High Power Expert Committee for estimating the investment requirements of urban infrastructure services, which stated that the Surat model should be replicated in other cities (High Powered Expert Committee 2011, Section 1.7.22, page 28). Consideration therefore needs to be given to how and whether the SCCT model can be applied in a suitable form in other cities, depending on their particular economic, demographic, and institutional contexts.

While the SCCT will require a degree of external support in its initial phase, it is increasingly ready to stand on its own two feet as it takes on the management and coordination of the end-to-end early warning system and the Surat Urban Health Centre. The experience of implementing these relatively soft projects will be invaluable in running other projects; linking multiple stakeholders; bidding for funding from donors, the private sector and government; and for pushing for the integration of climate change resilience considerations into all urban development plans and investment decisions for the city. The potential for the Trust to directly mobilise funds for city-level adaptation financing is crucial for sustaining future action, by decentralising financing mechanisms. The SCCT and associated climate initiatives therefore have the potential to change systems of urban governance in Surat, through improved mechanisms linking the relevant stakeholders to take not only a city-wide perspective, but also a region-wide approach, due to the trans-boundary nature of water management. In Indore, a more decentralised, sector-specific approach is being considered, whilst in Gorakhpur the approach is decentralised to the local ward level.

In parallel to this, at the national level, the mainstreaming of UCCR into urban development is being pushed from multiple angles. The need for training and capacity building at the city level has been recognised and the network of Indian institutions involved in UCCR activities will be able to apply their practical experience in advocating for policy measures, not only at the local scale but also with state and national bodies. While this mechanism is still in its early stages, through a two-pronged approach, there is now strong potential for lessons from local city-level experiences to inform top-down national directives in order to drive forward the climate adaptation agenda in Indian cities, in a way which also allows space for sharing of challenges and successes in building urban resilience. As per Fisher's (2012) conclusion, NGO networks – and by extension, networks which also include other actors such as educational bodies, such as the Consultation Committee – in India can play influential roles in designing and implementing development projects, all the more so in the emerging field of urban climate resilience.

Acknowledgements

This work was supported by the Rockefeller Foundation, as part of the Asian Cities Climate Change Resilience Network (ACCCRN) initiative.

References

Anguelovski, I. and J. Carmin. 2011. "Something Borrowed, Everything New – Current Opinion: Innovation and Institutionalisation in Urban Climate Governance." *Current Opinion in Environmental Sustainability* 3 (3): 169–175.

Bhat, G. K., A. Karanth, L. Dashora, and U. Rajasekar. 2013. "Addressing Flooding in the City of Surat Beyond its Boundaries." *Environment and Urbanization* 25 (2): 429–441. doi:10.1177/0956247813495002.

Brown, A., A. Dayal, and C. Rumbaitis del Rio. 2012. "From Practice to Theory: Emerging Lessons from Asia for Building Urban Climate Change Resilience." *Environment and Urbanization* 24 (2): 531–556.

Bulkeley, H., and V. Castán Broto. 2012. "Government by Experiment? Global Cities and the Governing of Climate Change." *Transactions of the Institute of British Geographers* 38 (3): 361–375. doi: 10.1111/j. 1475–5661.2012.00535.x.

Carmin, J., I. Anguelovski, and D. Roberts. 2012. "Urban Climate Adaptation in the Global South: Planning in an Emerging Policy Domain." *Journal of Planning Education and Research* 32 (1): 18–32.

Corfee-Morlot, J., I. Cochran, S. Hallegatte, and P. Teasdale. 2011. "Multilevel Risk Governance and Urban Adaptation Policy." *Climatic Change* 104 (1): 169–197.

Fisher, S. 2012. "Policy Storylines in Indian Climate Politics: Opening New Political Spaces?." *Environment and Planning C: Government and Policy* 30 (1): 109–127.

Gujarat State. 2006. *Gujarat Hazard Risk and Vulnerability Atlas 2006*. Gandhinagar: Gujarat State Disaster Management Authority.

High Powered Expert Committee (HPEC). 2011. *Report on Indian Urban Infrastructure and Services.* New Delhi: HPEC for Estimating the Investment Requirements for Urban Infrastructure Services. Accessed March 23, 2014. http://www.niua.org/projects/hpec/finalreport-hpec.pdf

Kernaghan, S. and J. da Silva. 2014. "Initiating and Sustaining Action: Experiences Building Resilience to Climate Change in Asian cities." *Urban Climate* 7 (2014): 47–63.

Leck, H. and D. Simon. 2013. "Fostering multiscalar collaboration and co-operation for effective governance of climate change adaptation." *Urban Studies*, 50 (6): 1221–1238.

Mani, N., and S. Wajih. 2014. A Participatory Approach to Micro-Resilience Planning by Community Institutions: The Case of Mahewa Ward in Gorakhpur City. Asian Cities Climate Resilience Working Paper, London: IIED, Number 7.

Ministry of Urban Development. 2013. *JNNURM Revised Toolkit For Preparation Of City Development Plan.* New Delhi: Ministry of Urban Development, Government of India. Accessed March 23, 2014. http://jnnurm.nic.in/wp-content/uploads/2013/05/CDP-Toolkit-Book-Second-Revised-2012-as-printing.pdf

McGranahan, G. et al. 2007. "The Rising Tide: Assessing the Risks of Climate Change and Human Settlements in Low Elevation Coastal Zones." *Environment and Urbanization* 19 (1): 17–37.

Revi, A. 2008. "Climate Change Risk: An Adaptation and Mitigation Agenda for Indian Cities." *Environment and Urbanization* 20 (1): 207–229.

Satterthwaite, D., S. Huq, H. Reid, M. Pelling, and P. Romero Lankao. 2007. "Adapting to Climate Change in Urban Areas: The Possibilities and Constraints in Low-and Middle-Income Nations." Human Settlements Discussion Paper. London: IIED.

Sharma, D., and S. Tomar. 2010. "Mainstreaming Climate Change Adaptation in Indian Cities." *Environment and Urbanization* 22 (2): 451–465.

Surat City Resilience Strategy. 2011. *Surat City Resilience Strategy.* Surat: ACCCRN, Surat Municipal Corporation, Southern Gujarat Chamber of Commerce & Industry, and TARU Leading Edge. Accessed March 23, 2014. http://www.acccrn.org/sites/default/files/documents/SuratCityResilienceStrategy_ACCCRN_01Apr2011_small_0.pdf

TERI. 2011. *Mainstreaming Urban Resilience Planning in Indian Cities, a Policy Perspective*. Delhi: The Energy and Resources Institute (TERI). Accessed March 23, 2014. http://www.acccrn.org/sites/default/files/documents/Final_Mainstreaming%20Urban%20Resilience%20Planning%20copy.pdf

Climate resilient planning in Bangladesh: a review of progress and early experiences of moving from planning to implementation

Neha Rai, Saleemul Huq and Muhammad Jahedul Huq

Bangladesh is one of the first least developed countries (LDCs) to develop a long-term climate change strategy, the Bangladesh Climate Change Strategy and Action Plan (BCCSAP). Two funds were set up after developing the BCCSAP, one using government resources (BCCTF) and the other using donor resources (BCCRF). This paper uses the "building blocks" framework to analyse changes that occur when progressing from planning to finance and implementation by comparing the BCCRF and BCCTF. This analysis reveals how governance enablers are influenced by political economy dynamics that steer funding decisions and implementation outcomes, and provides lessons for countries pursuing climate resilience.

Le Bangladesh est l'un des premiers pays les moins avancés (PMA) à avoir élaboré une stratégie à long terme en matière de changement climatique – Stratégie et Plan d'action du Bangladesh face au changement climatique (*Bangladesh Climate Change Strategy and Action Plan* – BCCSAP). Deux fonds ont été mis en place après la création de la BCCSAP, un utilisant des ressources gouvernementales (BCCTF) et l'autre utilisant des ressources provenant de bailleurs de fonds (BCCRF). Cet article se sert du cadre « *building blocks* » (éléments constituants) pour analyser les changements qui surviennent lors du passage de la planification au financement, puis à la mise en œuvre, en comparant le BCCRF et le BCCTF. Cette analyse révèle en quoi les facilitateurs de la gouvernance sont influencés par la dynamique de l'économie politique qui oriente les décisions de financement et les résultats de la mise en œuvre, et propose des enseignements à l'intention des pays en quête de résilience au changement climatique.

Bangladesh es uno de los primeros países de menor desarrollo (LDC) que ha diseñado una estrategia de largo plazo para hacer frente al cambio climático: el Plan estratégico y programático ante el cambio climático de Bangladesh (BCCSAP). Tras su elaboración, se crearon dos fondos, uno financiado por el gobierno (BCCTF) y otro apoyado por donantes (BCCRF). A través de la comparación de ambos fondos, el presente artículo utiliza el marco de los "bloques de construcción" para analizar los cambios que tienen lugar cuando se avanza desde la planificación hacia la financiación y desde ahí a la implementación. Dicho análisis revela la forma en que los facilitadores de la gobernanza acusan la influencia de las dinámicas de la economía política, las cuales orientan las decisiones de financiamiento y los resultados obtenidos tras la implementación, brindando, además, aprendizajes útiles a aquellos países en los que existe interés en promover la resiliencia ante el cambio climático.

CLIMATE CHANGE ADAPTATION AND DEVELOPMENT

Introduction

Bangladesh is often cited as one of the most climate vulnerable countries in the world due to its geophysical location, exposure to extreme conditions caused by climatic stimuli, and high population growth rate (Huq and Ayers 2007; Parry et al. 2007; GoB 2009a, 2009b). Additionally, although the GDP of the country is consistently showing progress, one third of the population lives below the poverty line (MOF 2013). High population density and dependency on agriculture and natural-environment based livelihoods have further increased vulnerabilities over the years due to hazards and uncertain climatic variability. The combinations of these stresses make Bangladesh one of the most affected least developed countries (LDCs) when it comes to climate change-related issues. According to the Ministry of Environment and Forests (MoEF 2012), increased frequency and intensity of climate-induced natural disasters account for losses worth 1.5% of the country's GDP. A study conducted by the Planning Commission of the Government of Bangladesh (GoB) highlights that changes in crop production due to climate change increase the number of poor people and are equal to the rate of total crop damage caused by any major disaster (GoB 2009c). Dasgupta et al. (2011) estimate that it could cost US$2671 million by 2050 to protect the major towns of Bangladesh from climate-induced monsoon flooding (Dasgupta et al. 2011). Further research reveals that on average, 57% of annual development investment is now at risk of being adversely affected by climate change and an additional 10–30% of funding is required to ensure the current level of benefits from development projects (Haque 2009).

These inextricable linkages between climate change and development make it critically important to synergise climate and development policy actions to support adaptation. Although self-evident, in reality, mainstreaming climate change adaptation into development activities remains a challenge (Adger et al. 2007). Burton (2004) recognises this as an "*adaptation deficit*", defined as a "*failure to adapt adequately to existing and future climate risks*" thereby causing a discrepancy between existing and optimum levels of adaptation. Climate disjointed decisions made by development practitioners will further increase the deficit (Burton 2004). Bridging the adaptation deficit requires measures to be integrated into coherent development strategies. This in turn requires improved governance and mainstreaming of climate change measures into national development objectives (Davidson et al. 2003). Although still in its early years of institutional restructuring, Bangladesh, amidst a range of constraints, has made remarkable progress in climate resilient development planning over recent years. It is also one of the most proactive countries in global climate negotiations, as well as in addressing its own climate issues. The advanced position of Bangladesh as a global leader in adaptation is in part due to strong political will and an enabling environment, encouraging the mainstreaming of climate risk management at different levels of planning and implementation throughout country governance. Recognising this comparatively advanced position of Bangladesh in addressing its adaptation deficit, practical actions, and responses to climate change therefore provides interesting lessons that can inform decision-making for climate change-related efforts across countries facing similar challenges. This paper applies the International Institute of Environment and Development's (IIED) "building blocks framework for mainstreaming" (Figure 1) to understand the existing support systems for climate resilient development planning in Bangladesh, including the enabling environment, policies, and institutional and financial frameworks.[1] This framework, although an effective tool to characterise a country's status and trends in the evolution of climate mainstreaming efforts, offers limited scope to assess the effectiveness of these building blocks. To understand the strength and outcomes of the building blocks, a deeper political economy analysis is needed to attribute the outcomes, which is not entirely within the scope of this paper. Nonetheless, we highlight initial governance reflections that direct us to further research in this area.

The building blocks framework identifies three building blocks for successful climate change mainstreaming into development planning: (1) an enabling environment; (2) policies and

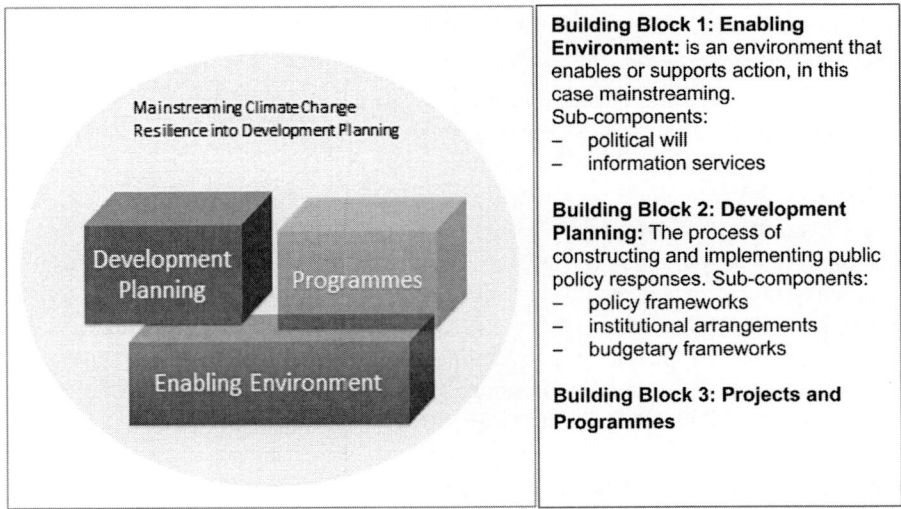

Figure 1.   Building blocks for mainstreaming climate resilience into development planning.
Source: Pervin et al. 2013.

planning; (3) projects and programmes (as illustrated in Figure 2). By applying this framework to the case of Bangladesh it is possible to identify the combination of blocks that has enabled the government to incorporate adaptation into planning and implementation processes. The paper traces changes in the country's progress from planning to implementation by comparing the embedded cases of two main funding instruments in Bangladesh: the Bangladesh Climate Change Trust Fund (BCCTF) and Bangladesh Climate Change Resilience Fund (BCCRF). Through this analysis it is possible to identify lessons that are relevant to countries working towards developing plans and institutions for climate change, as well as those moving from planning to implementation.

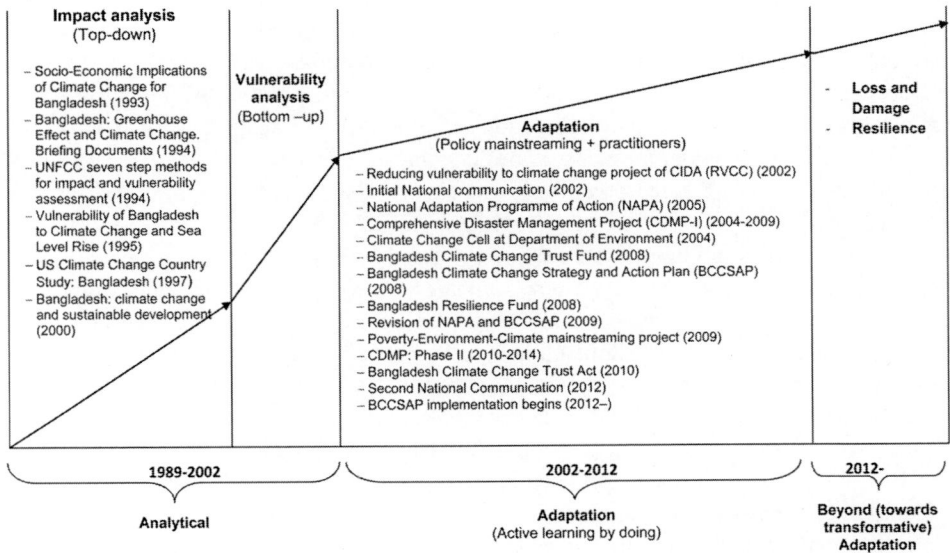

Figure 2.   Evolution of climate change adaptation in Bangladesh.

How is Bangladesh planning to address its adaptation deficit? Key building blocks

Proactive climate-resilient planning: policy and institutional architecture

The discourse around climate change implications on development has evolved over the last decade, and is now a critical part of policy agendas and international climate change discourse (Schipper 2006). In 2002, CARE-Bangladesh implemented the country's first adaptation project ("Reducing Vulnerability to Climate Change") using a community-based approach. In 2005, the Government of Bangladesh (GoB) prepared its National Adaptation Programme of Action (NAPA) in line with international policy requirements, identifying 15 urgent and immediate adaptation actions. Later, the Bangladesh Climate Change Strategy and Action Plan (BCCSAP), as an extension of NAPA, was prepared in 2008 following the Bali Action Plan. The BCCSAP is a 10-year plan aiming to facilitate medium and long term adaptation measures, including 120 projects that are captured under six thematic areas, in order to encourage low-carbon and climate resilient development (GoB 2009b). These thematic areas are: (1) Food security, social protection and health; (2) Comprehensive disaster management; (3) Infrastructure development; (4) Research and knowledge management; (5) Mitigation and low carbon development; and (6) Capacity building and institutional development. National planning documents identify the adverse impacts of climate change as one of the crucial and evolving challenges facing Bangladesh and therefore focus on mainstreaming climate change adaptation into sectoral policies, plans, and programmes. Figure 2 represents the key efforts to address the adaptation deficit that underlines a gradual evolution of climate change adaptation in Bangladesh.

As illustrated in Figure 2, the government is steadily increasing its efforts to establish policies, institutions, and plans in order to nurture and enhance the adaptive capacity of the country and increase resilience to climate change risks. An enabling environment; the policy, institutional, and financial arrangements; various projects and programmes for addressing climate risks – all constitute the essential "building blocks" needed for successful mainstreaming of climate change. Bangladesh has nurtured these building blocks over the last few years, emerging as an exemplary arrangement for addressing climate risks.

Building Block A: An enabling environment for mainstreaming includes the political will to make climate policy and the information services that guide it. The *political will* for addressing climate issues is evident in the election manifesto of the ruling party of Bangladesh, the Awami League, with a focus on climate change as an environmental and developmental priority (Pervin 2013). The widespread awareness of climate change amongst voters and civil society in Bangladesh has made it an important political issue for any ruling party. A Cabinet Review Committee has also been set up to review the actions of Bangladesh's main strategy and action plan, led by the prime minister under the Ministry of Planning. The formation of this plan and allocation of the government's own budget towards climate change also reflect the political will within the country. This is in contrast to many developing countries, where development and adaptation planning and finance could be primarily driven by international agendas.

Building Block B: Development planning includes the policy frameworks together with institutional arrangements and finance mechanisms. The policy frameworks, strategies and action plans for Bangladesh clearly reflect the climate resilience[2] objectives of the country. Bangladesh was the first country to develop a National Adaptation Programme of Action in 2005, which identified 15 urgent and immediate adaptation actions. NAPA was updated in 2009, directly translating into Bangladesh's first strategy on climate change, the Bangladesh Climate Strategy and Action Plan (BCSSAP). Climate resilience is also reflected throughout policy *planning cycles*. The Planning Commission in Bangladesh has begun a process to internalise climate change into long and medium term planning processes. Climate change is also reflected in the sixth development plan and the national perspective plan of Bangladesh. These efforts to mainstream climate change through governance processes indicate that the political rhetoric for climate change is actually being translated into concrete steps towards building adaptive capacity.

Building Block C: Institutional frameworks for climate change are also established within various ministries, departments and agencies. The Ministry of Environment and Forests (MoEF) is the key agency responsible for climate change-related matters. Several other institutions have evolved since the preparation of the BCCSAP, including a Management and Technical Committee for the Bangladesh Climate Change Resilience Fund (BCCRF) and Trust Fund (BCCTF). Climate Change Cells (CCC) and a Climate Change Unit (CCU) have also been established to coordinate ministries, and a Department of Climate Change is also being developed. The CCC's are the longer arms of the Comprehensive Disaster Management Programme (CDMP) housed in the Department of Environment, which provides technical support on climate change-related issues to MoEF. A Climate Change Trust (CCT) was established in the MoEF to coordinate and manage the climate action plan, later assimilating into the BCCTF. Climate change focal points in all ministries are formed to liaise with the MoEF and to support smoother implementation of the climate action plan. Currently these focal points are not very active or functional and major decision-making rests with the MoEF. However, there is a National Environment Committee in the country headed by the prime minister for strategic guidance and oversight, and a National Steering Committee on climate change chaired by the Minister of MoEF to harmonise the progress of all climate related activities in Bangladesh. The establishment of institutions further signifies commitment to developing the longer term capacity of the government to address climate change adaptation and promote resilience.

Building Block D: Finance mechanisms for climate change efforts in Bangladesh were first set up in 2002 with support from the Canadian International Development Agency (CIDA) to implement the "Reducing Vulnerability to Climate Change" (RVCC) project. This was subsequently followed by the creation of a national "Climate Change Trust Fund" (BCCTF) in 2008 to support the implementation of the BCCSAP. Bangladesh has become one of the first LDCs to allocate US$350 million from its non-development budget towards climate risk management (Pervin 2013).

The BCCTF has been operationalised through government approval of a Climate Change Trust Fund Act passed in 2010. The BCCTF receives a block budgetary allocation of US$100 million per year from GoB towards climate change (Figure 3). Apart from the BCCTF,

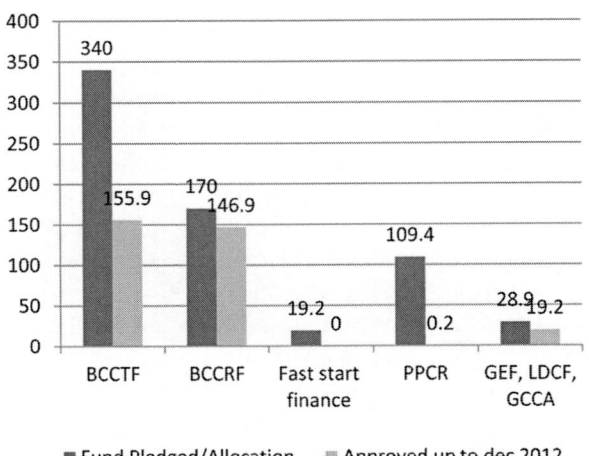

Figure 3.   Financing windows for climate change in Bangladesh.

Bangladesh has several other institutional funding mechanisms for climate change adaptation. The Annual Development Programme, which is led by the Planning Commission and Ministry of Finance, allocates up to 4% of GDP to delivering climate response actions (Pervin 2013).

Major bilateral and multilateral development partners also promote climate change as a central development issue (Alam et al. 2011; ICAI 2011; SDC 2010) and have set up a multi-donor fund to implement the BCCSAP and "climate-proof development" in Bangladesh (Ayers 2009, 239). The BCCRF is one such donor-funded arrangement, where the World Bank acts as an interim secretariat, playing a key role in facilitating BCCRF in due diligence of projects, fiduciary management, ensuring transparency and accountability. Additionally, the Pilot Programme for Climate Resilience (PPCR), an adaptation fund under the Climate Investment Fund (CIF) of the World Bank, supports mainstreaming of climate risk and resilience into the country's core development planning and implementation (CIF 2011). Bangladesh also receives support from the Least Developed Country Fund (LDCF) through the Global Environmental Facility (GEF) to implement its NAPA. These multi-lateral funding arrangements are parallel to additional bilateral development assistance.

Bilateral development cooperation has led to the direct implementation of different climate change-related projects in Bangladesh (Hedger 2011).These bilateral initiatives jointly support different types of mainstreaming activities setting the country on its adaptation path. The modalities and mechanisms employed by different funding windows vary. Support for the national budget is seen as one way of promoting more integrated climate resilience, whereas the BCCRF and Climate Investment Funds are observed to be more stand-alone pilot projects intended to meet specific sub-objectives of the BCCSAP (MoP 2012; Pervin 2013). In reality, the former also suffers from its inherent management weaknesses. Nonetheless, these funding innovations place Bangladesh in a strong position to respond to and address climate risks. The large amounts of multilateral and bilateral financial support also provide strong incentives for the government to continue promoting climate change adaptation as a political and developmental imperative.

Moving from planning to practice

Since its formulation in 2009, the objectives of BCCSAP are being supported by various funding windows and initiatives. Government agencies and non-governmental organisations are pursuing a mixture of adaptation projects guided by the BCCSAP. A summary of these adaptation activities and their major funding sources is given in Table 1.

According to the BCCSAP, climate change adaptation is defined as "climate-proofing" development and reducing the impact of climate change on economic growth and poverty reduction. The BCCSAP was prepared by identifying cause and effect relationships (Figure 4), which are based on the impact that past extreme events had on development in Bangladesh, and used to identify climate vulnerable sectors and geographies.

Based on this analysis of vulnerabilities, the BCCSAP proposes a list of generic measures (Table 2) to address the likely impacts of global warming on Bangladesh and help the government secure external support to implement those activities.

Although a range of measures were prioritised, it is important to note that many of the measures represent a relabeling of older concepts of development that are translated into different local contexts and give greater consideration to spatial, temporal, and structural dynamics (e.g., innovative flood management or drought and saline tolerant crop varieties). These measures range from continuation of current best practices to the substantial enhancement of existing technical approaches, such as improved irrigation and up-gradation of flooding and coastal embankments (Alam et al. 2011).

Table 1.   Sources of different adaptation funding and their preferred adaptation actions.

	Bangladesh Climate change Trust Fund (BCCTF)	Bangladesh Climate Change Resilience Fund (BCCRF)	Pilot Programme for Climate Resilience (PPCR)	Global Environmental Facility (GEF)
Contributing amount	US$100 million annually	US$188.2 million	US$110 million	US$9.65 million (July 2012-June 2014)
Type of contribution	–	Grant	US$50 million as grant + US$60 million as loan	Grant
Funding source	Annual revenue budget of Bangladesh	Contribution from DFID (UK), SIDA (Sweden), DANIDA (Denmark), USAID (USA), EU, SDC (Switzerland)	Climate investment fund	Least Developed Countries Fund (LDCF)
Implementing organisations	Different ministries and national NGOs through PKSF	Different ministries and PKSF	Different ministries, implementing MDBs	UNDP Bangladesh MoEF and NGOs
Infrastructure (cyclone shelter, flood shelter, embankment, river training and dredging, and road construction.	√	√	√	
Technical assistance		√	√	
Community based adaptation project	√	√ (10% of total fund)		√
Coastal afforestation and reforestation (Coastal green belt)	√	√	√	√
Insurance				√

Sources: MoEF 2010; MoEF 2013b; CIF 2010; GEF 2012.

A closer look at the thematic focus of the BCSSAP as well as the nature and approaches to climate change adaptation helps us to understand how climate resilient planning is translated into actions and practices. Looking at BCCSAP proposed adaptation activities in six thematic areas, one can arrive at two broad classifications: (1) infrastructure based adaptation, and (2) community-based adaptation. *Infrastructure based adaptation* requires state involvement for large investments, whereas *community-based adaptation* evolves from the country's

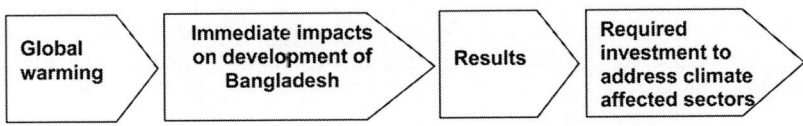

Figure 4.   BCCSAP used cause-effects relationship in the context of global warming and development in Bangladesh.

Table 2. Generic measures proposed to increase the adaptive capacity of vulnerable communities.

- Early warning system
- Cyclone shelters and *killas*
- Improved operation and maintenance of (coastal) embankments and polders
- Upgrading of flood protection embankments/drainage systems
- Raising roads and railway tracks
- Flood proofing
- Improved crop and cropping system
- Improved irrigation and water management
- Provision of (potable) drinking water and sanitation
- Possible industrial relocation
- Health education/awareness and immunisation
- Social protection as a crosscutting issue

traditional community-based development approaches. Community-based adaptation is shown in the BCCSAP as a spin-off effect of social protection. These focus areas are further explored under the two funding instruments: BCCRF and BCCTF.

From planning to implementation: the government-funded Bangladesh Climate Change Trust Fund

According to the Climate Change Trust Fund Act 2010 (MoEF 2010), climate change adaptation projects under the BCCTF can only finance public sector and national NGOs through a competitive process, and the duration of the projects vary from one to two years. The Climate Change Trust Fund Act offers three different tiers to operationalise the fund and ensure transparency and accountability in its disbursement system. The Board of Trustees is responsible for the overall governance and management of the trust fund and consists of 17 members, including two from civil society organisations nominated by the GoB. This trustee board is also the apex decision-making body that approves projects for the Trust Fund. It is chaired by the Minister of MoEF. The technical committee, headed by the Secretary of the MoEF, reviews project proposals, develops annual work plans and budgets for the trust, and helps the Board of Trustees develop policy and make funding decisions. This technical committee has 12 members from different ministries and two sub-technical committees (ecosystem and technical) staffed by experts to provide particular technical advice to the technical committee. The Climate Change Trust acts as a secretariat and is held accountable for all proposals, preliminary project proposal screening, fund disbursement, and monitoring and evaluation of the ongoing projects, whereas the MoEF is a coordinating agency for policy execution. As per a new order within the Board of Trustees, the Palli Karma Soyahok Foundation (PKSF), which is the apex body of microcredit organisations in the country, has been given accountability for overseeing the off-budget part of this fund (NGO window), to build capacity and awareness on climate change, and support community level climate change adaptation projects.

The governance mechanism defined in Figure 5 gives a picture of how climate resilience is being mainstreamed into different implementing ministries as well as civil society. As per Climate Change Trust Fund policy 2010 (MoEF 2010), 66% of that total fund will be spent on executing the prioritised actions and programmes of the BCCSAP and the remaining 34% will be put into a fixed deposit in a bank for national emergencies. The interest of that fixed deposit will also be utilised for BCCSAP projects.

As of September 2013, the Government of Bangladesh has selected 139 projects worth US$200 million for implementation by different government departments and NGOs (MoEF

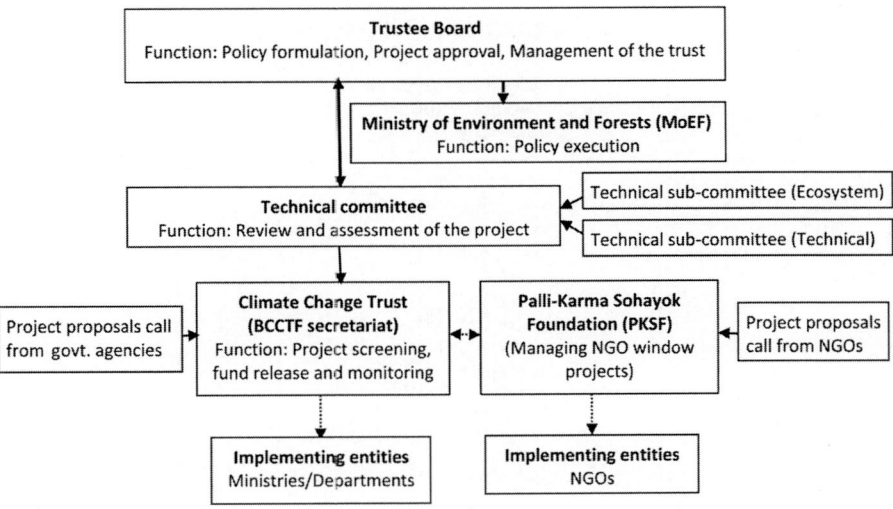

Figure 5.   Governance mechanism of Bangladesh Climate Change Trust Fund.

2013a). Although the trust fund is being utilised in sync with the development priorities of the BCCSAP, one can observe a strong preference towards specific themes and practices (Figure 6). For example, two-thirds of the total allocated funding is designated for *water infrastructure* in coastal areas. Water infrastructure is followed in importance by mitigation and low carbon development (20.21%). Food security, social protection, and health, and comprehensive disaster management receive 8.05% and 5.83% respectively. The other two thematic areas, research and knowledge management and capacity building and institutional strengthening, receive 3% of total funding each. Although adaptation continues to remain in the heart of Bangladesh's priorities, actions under BCCSAP's fifth theme, low carbon mitigation, are gaining momentum in the implementation plans of BCCTF (although less so when compared to the BCCRF). Some of these planned mitigation projects comprise of solar electrification, recycling, solar irrigation, and afforestation projects (GoB 2014).

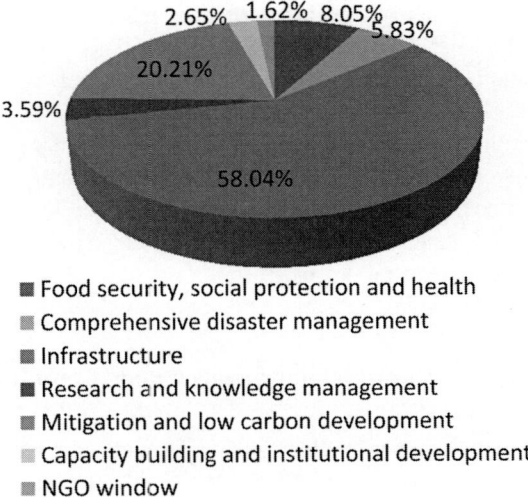

Figure 6.   BCCTF allocation under BCCSAP's six thematic areas.

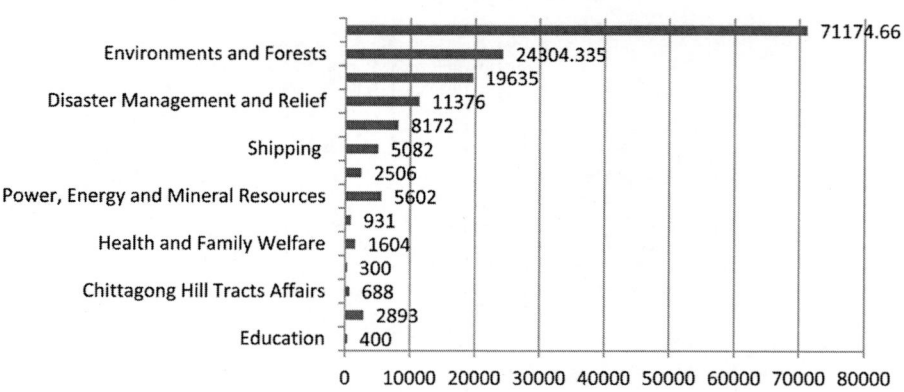

Figure 7.   Funding to ministries (in Bangladesh Taka-BDT).

Amongst the adaptation priorities, some of the infrastructure projects which received early approval were in coastal areas. These include construction of cyclone resilient houses, construction of dams, and coastal afforestation projects. Although the first set of projects received quick approvals, some received criticism for neglecting safeguards and compliance to proper planning procedures. For example, one of the cross dam projects approved by the trust fund was criticised for its inadequate social and environmental impact assessments. Lack of coordination between ministerial departments further weakened the case; the project stalled when the Forest Department opposed the proposal of a dam crossing through a reserve forest (TIB 2014). Although a coastal geographical focus and targeting of infrastructure are both political and strategic priority areas and issues in Bangladesh, the unconvincing process of prioritisation and implementation have brought the system under criticism. Some areas of criticism include inadequate openness about the project selection mechanisms (TIB 2013).

Apart from project selection issues, the allocations of the six climate funds have been concentrated within a selective few ministries and departments. Indeed, the Bangladesh Water Development Board (BWDB), the government body under the Ministry of Water Resources that is responsible for constructing water-related infrastructure, receives the highest amount (45%) of funding from the Trust Fund (58 projects) (MoEF 2013a). This diverges from the BCCSAP's vision to bridge the adaptation deficit by integrating climate resilient actions across the activities of line ministries (Figure 7).

Although BCCTF continues to be an example of country leadership in responding to climate change, initial reflections on governance issues explored above point to the risks when moving from planning to implementation. The fund has come under scrutiny for partisan politics and exceptionalist planning (illuminated by Alam et al. 2011) clearly showing that implementation decisions are not necessarily linear processes.

From planning to implementation: the donor-financed Bangladesh climate change resilience fund

The BCCRF uses a similar governance mechanism and the same competitive process for selecting and prioritising adaptation projects as the BCCTF. It also uses the same implementing government institutions and non-state actors, including Bangladeshi NGOs, civil society organisations, community-based organisations, research institutions, and other civil groups. The major

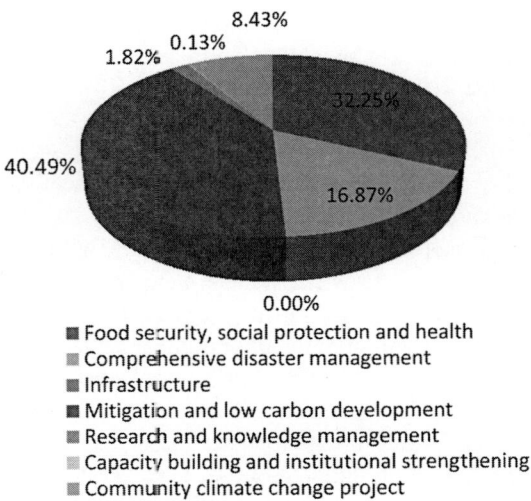

8.43%
0.13%
1.82%
32.25%
40.49%
16.87%
0.00%

■ Food security, social protection and health
▧ Comprehensive disaster management
■ Infrastructure
■ Mitigation and low carbon development
▧ Research and knowledge management
▨ Capacity building and institutional strengthening
▧ Community climate change project

Figure 8.   Distribution of BCCRF under thematic areas of BCCSAP.

difference between the BCCTF and BCCRF is the membership of the technical committee, its management by the World Bank, and the presence of international donors on the governing council. Indeed, the World Bank plays a key role in assisting the BCCRF to screen and supervise projects.

Grant criteria for BCCRF funding include the requirements that projects should be between US$15 and US$25 million, be demand driven, and have three-year timelines with a possible one-year extension. Additionally, proposals coming from existing project units are given advanced consideration during selection and prioritisation.

The distribution of BCCRF funding to the thematic areas of the BCCSAP is given in Figure 8. Although adaptation continues to be at the forefront of BCCRF priorities, projects under the mitigation and low carbon development theme of the BCCSAP now constitute almost 40% of the total funding. Indeed, one of the largest projects funded under the BCCRF is a mitigation project funding solar-powered irrigation pumps, costing US$60 million; it will be funded by the BCCRF, the World Bank, and the Infrastructure Development Company Limited (IDCOL). The project aims to decrease Bangladesh's reliance on foreign diesel and improve food security (GoB 2014).

A major part of the adaptation pool is going into three key programmes: (1) *building multi-purpose cyclone shelters in coastal areas*; (2) *agriculture adaptation in climate risk prone areas of Bangladesh – drought, flood, and saline prone areas*; (3) *afforestation and reforestation for climate change risk reduction in hilly and coastal areas*. Additionally, the BCCRF is supporting six studies to inform the design of future adaptation projects in various sectors. The NGO window within the BCCRF receives 10% of this total fund, which is managed and implemented by PKSF. PKSF is managing a community level climate change project following an agreement in August 2013 to fund selected NGOs.

As of September 2013, donor agencies contributed US$188.2 million to the BCCRF and 81% of this was already disbursed to 13 projects, including the NGO window. Apart from this NGO window, six analytical studies are funded under the thematic area of research and knowledge management, and there are an additional five investment projects. Of the five investment projects, three are stand-alone projects and the rest are co-financed with other ongoing World Bank projects. There is also a project funded under the thematic area of capacity building and institutional strengthening that intends to establish the BCCRF secretariat as a key mechanism for attracting climate finance into Bangladesh. The BCCRF is thus supporting a comprehensive programme of work with a balanced combination of adaptation approaches including infrastructure, research, and knowledge management.

Although it is still too early to measure the development impact of these approaches, public expenditure in disaster management has proven to be effective. Better infrastructure and information systems were both credited with reducing the number of fatalities that would have otherwise been caused by the Sidr cyclone in 2008 (Haque et al. 2012). Building on these past successes, the BCCRF has scaled up construction of multipurpose shelters and better early warning systems. These newly constructed shelters were highly utilised during the Mahsen cyclone in 2013. Cyclone shelters, approach roads, and *killas* (storage shelters) all played a vital role when Mahsen forced more than one million inhabitants in 13 coastal districts to take shelter (MDMR 2013). An enhanced early warning system further enabled inhabitants to respond and seek shelter in a timely manner. Early warning systems are also employing a public-private partnership model that brings together the government and mobile providers to supply early warning information through text messages (ICAI 2011).

Apart from investments in targeted adaptation mechanisms, BCCRF funding towards knowledge management, capacity building, and institutional strengthening is also building management capabilities and stewardship of national entities. This is evident from the fact that disaster risk reduction and climate risk management are being included in the work and plans of eight ministries (ICAI 2011). The role of the World Bank in the fund also brings in technical value, consideration of safeguards, reduced corruption, and compliance to fiduciary standards. However, the management of the BCCRF by external entities is of concern to the government of Bangladesh, who had expected complete fund ownership by the government once their management capacities were built. The BCCRF is challenged by differences between country and development partner priorities, which increasingly raise concerns about country ownership.

The focus on adaptation has been a specific policy agenda for the Bangladeshi government for some time, leading the emerging emphasis on low carbon projects to be contested as an externally driven agenda by campaigning groups in the country (Alam et al. 2011).

Despite this, there is a growing acceptance that mitigation will have a role in Bangladesh's climate change strategy, not least because the Government of Bangladesh feels it is necessary in order to continue to secure foreign aid and is increasingly concerned about energy security. It is worth emphasising that mitigation and adaptation are not necessarily differing policy aims. Numerous afforestation projects funded by the BCCRF and the BCCTF achieve both mitigation and adaptation objectives, by reducing the risk of flooding as well as acting as carbon sink. Although all of the projects have been divided into six thematic areas, success will mean that the benefits of many will be felt well beyond their own, specific thematic area.

Early lessons from planning to implementation: some results and comparison of the BCCTF and BCCRF

The BCCSAP sets out a range of adaptation options and priorities to be implemented. Most programmes and projects are in the early design phase; but they clearly intend to address the development needs of society while building climate resilience. Planned projects include addressing chronic disaster vulnerabilities and food security in order to build the adaptive capacity of climate vulnerable communities. This focus also extends to the infrastructure thematic area, where respective government agencies place emphasis on building shelters, early warning systems, excavating canals, and constructing infrastructure to manage increased flooding due to climate change, as well as protect riverbanks from erosion in order to protect settlements.

Most interventions funded by the BCCTF and BCCRF are in the early stages and it is difficult to know the extent to which investments have resulted in reduced vulnerability, which may take decades to accurately understand. However, past experiences show that scaling up of climate specific interventions (e.g., DRR related infrastructure) *does* reduce the vulnerability of coastal communities to cyclones, albeit indirectly.[3] Both the BCCTF and BCCRF have selected and prioritised climate change responses that have proven to be effective in the past. Investments made in coastal Bangladesh are a good example, as early warning systems, cyclone shelters, and capacity building in disaster management have all proven to significantly reduce disaster-induced death rates and protect human assets. A comparison of cyclones in 1991 and 2007 also reveals how infrastructure development has helped to reduce death rates and vulnerability in the region (Paul 2009). Indeed, investments in infrastructure capacity and better coordination kept the death toll of cyclone Sidr (2007) below 3000 compared to a similar category (IV) cyclone in 1991 which had a death toll of more than 100,000. In Myanmar in 2008, a country which is very similar to Bangladesh in all aspects except disaster preparations, Cyclone Nargis caused 140,000 deaths and affected around 2.4 million people (Lateef 2009). The case of Myanmar shows how poor dissemination of information, inadequate early warning systems, poor government coordination, and inadequate infrastructure can all intensify the impacts of natural disasters (Haque et al. 2012). In comparison to Myanmar, Bangladesh experienced decreased fatality rates (e.g., Sidr in 2007, and Aila in 2009) because of modernised early warning systems, as well as effective use of infrastructure and coordinated government efforts. It is perhaps understandable that scaling up of similar investments was a preferred option by GoB.

Although it is still too early to see the developmental impacts of these adaptation investments, we can say that it has resulted in increased awareness and knowledge amongst different stake-holders engaged in climate risk management. An overview of policies, plans, funding mechanisms, and their institutionalisation within development projects provides some evidence that building blocks for climate risk management are being put in place and stakeholder knowledge is increasing. In order to understand how these building blocks shape funding decisions and continued implementation, it is helpful to compare the two main funding and implementation mechanisms, the BCCTF and BCCRF.

Building Block 1: The enabling environment for implementation under the BCCTF and BCCRF appears to be heavily shaped by the composition of the governance and technical boards. Furthermore, the composition of these boards is shaped by those providing financing, which is evident by the stronger influence of development partners in the BCCRF. The establishment of the BCCTF using national government resources is a noteworthy achievement for Bangladesh and may allow greater ownership in implementing adaptation activities throughout the country. In this manner, although the BCCSAP is being used to inform the funding and implementation decisions of both funds, there are some emerging challenges that need consideration in future planning and implementation (for example, differing transparency and accountability mechanisms).

Building Block 2: Development planning also appears to be heavily shaped by the governance and technical composition of the BCCTF and BCCRF. The BCSSAP has laid out 44 priorities within six themes drawing from the NAPA priorities. However, decision makers within the funds appear to cherry pick specific projects and actions amongst the wide range of priorities for making funding allocations. The BCSSAP is clearly a conceptual strategy, but because there is no strict implementation strategy the plan is reinterpreted depending on the context and fund (Khan, Huq, and Shamsuddoha 2012). The BCCTF has already given birth to controversies about the transparency and accountability of its project selections and funding allocations

(Khan, Huq, and Shamsuddoha 2012). Political bias has significantly influenced project selection and funding decisions (anonymous interview with a MoEF official). In addition, the BCCTF has no specific project selection and prioritisation indicators or criteria. The BCCRF is also considered to diverge from principles of ownership due to its management by an external agency. Its emerging emphasis on low carbon development has also brought it under civil society scrutiny (Alam et al. 2011).

Finally, although both the BCCTF and BCCRF aspire to mainstream climate change adaptation activities in the country's national development planning process, the bulk of funding by these funds is going towards a few lead line ministries without wider representation of national interests. The Ministry of Planning has a limited role in the funds' governance mechanisms. The approval process of BCCRF projects is also criticised for bypassing the normal planning process of the country. The Ministry of Planning argues that since these adaptation projects are not prepared following national planning development guidelines, they have no opportunity to be mainstreamed into the national annual development planning of Bangladesh. In response it is argued that the customary development planning procedure of Bangladesh has been avoided due to its lengthy process and to ensure the quick utilisation of the fund.

Building Block 3: The nature of projects and programmes used to implement the BCCSAP are strongly shaped by the enabling environment and development planning that takes place within each fund. The selected projects under both BCCTF and BCCRF depict an uneven distribution of funding in terms of BCCSAP thematic areas. Coastal areas are agreed to be one of the hotspots in terms of climatic vulnerability; it is noticeable that both the funds have concentrated in coastal areas of the country, without addressing the vulnerability issues of other hotspots. This indicates a dilemma in appropriate project selection and prioritisation. Additionally, the overemphasis on infrastructure development makes us question the selection process in relation to targeting the poor – as development outcomes of infrastructure-based risk reduction projects often benefit those at the top of the income ladder while ignoring the rural communities at risk. The disbursement of finance within the NGO window also raises controversy around the selection of NGOs based on the personal affiliations of stakeholders, when political interest has significant influence over project and partner selection. Although the BCCRF benefits from the World Bank's administrative systems for maintaining fiduciary standards, these are also delaying the implementation of projects. The BCCRF is often criticised for its slow bureaucratic process in project selection and fund spending. Most of this fund is tagged as "additional" to other World Bank projects in Bangladesh.

Conclusions

An initial reflection on Bangladesh's experience in planning and implementing climate change responses highlights early lessons learnt that have importance for developing countries seeking to follow a similar development path. Bangladesh's experience of the BCCTF and BCCRF shows how important it is to reflect, nurture, and steer the building blocks needed to complete each stage leading up to implementation. This may mean understanding the political economy of a system in order to steer that system toward a common development path.

Perhaps we can draw some lessons from these initial findings and their implications on mainstreaming and ensuring transparent outcomes. For example, it is worth reiterating that although both funds have set up systematic ways to allocate funding to Bangladesh's priority themes, transparency in decision making has come under scrutiny, particularly in the case of the BCCTF. Selective prioritisation during planning and low levels of coordination between parallel funding mechanisms can run the risk of leaving gaps and increasing the "adaptation deficit". A better understanding of political interests and influence early on can help assess risks in funding decisions (for example, biases in project selection or neglect of safeguards). This may mean including independent entities for screening and overseeing actions or integrating better safeguard mechanisms.

A further reflection on the building blocks shows that the issue of climate change is being successfully mainstreamed within the policies, plans, institutions, and programmes of Bangladesh. However, allocation of large amounts of funds to a few key lead ministries is a sign of selective mainstreaming toward ministries that are targeting finance to relatively manageable and also politically important areas and issues. Mainstreaming climate resilience goes beyond implementation of projects by a few key actors; wider involvement is required to ensure that effective planning reflects wider national interests.

As countries move from planning to practice, these early experiences reveal that planning and implementation decisions are much more than just linear, simple outcomes. Implementation decisions are complex and do not necessarily progress in a linear fashion. Besides policy and institutional drivers; actor behaviours, underlying incentives, interests, and partisan politics influence actions and decision-making. These are the political economy dynamics that often challenge good governance and strongly influence implementation outcomes. In this paper we apply the building blocks framework to analyse how Bangladesh is nurturing its policy and institutions to address and respond to the issues of climate change. The framework is an effective way to characterise and understand a country's status and trends in evolution of climate mainstreaming efforts. Bangladesh appears to have nurtured its policies and institutions to respond to climate change; however, setting up building blocks alone is not necessarily a reflection of how well they function or perform. Various actor interests and incentives influence the outcomes of the systems which are put in place. A better understanding of the political economy dynamics early on can be used as an effective way to steer the building blocks towards achieving better development impacts. This could mean mapping interest, influence, and commitment of relevant actors prior to establishing a financial mechanism that will have major implications for steering decisions that prioritise the poor. Although this paper sets out some initial reflections on politics and governance issues while countries move from planning to implementation, a deeper political economy analysis is needed to analyse and attribute the outcomes to climate change responses.

Notes

1.   The Building Blocks framework was developed by a diverse group of government staff from least developed countries (LDCs) who came together at a course facilitated by IIED and ICCAD to share and reflect on their country's experiences and needs around integrating climate change into development planning.
2.   Climate change resilience is the ability and the capacity of the system to maintain its integrity and recover from a situation when subject to climate change-related disturbances.
3.   Vulnerability measured in terms of proxy indicator 'reduced death rate'.

References

Adger, W. N., S. Agrawala, M. M. Q. Mirza, C. Conde, K. O'Brien, J. Pulhin, R. Pulwarty, B. Smit, and K. Takahashi. 2007. "Assessment of Adaptation Practices, Options, Constraints and Capacity. Climate Change 2007: Impacts, Adaptation and Vulnerability." Contribution of Working Group II to the Fourth Assessment Report of the Intergovernmental Panel on Climate Change. Cambridge: Cambridge University Press.

Alam, K., M. Shamsuddoha, T. Tanner, M. Sultana, M. J. Huq, and S. S. Kabir. 2011. *Planning Exceptionalism? Political Economy of Climate Resilient Development in Bangladesh*. UK: Institute of Development Studies (IDS). *IDS Bulletin*. East Sussex.

Ayers, J. 2009. International Funding to Support Urban Adaptation to Climate Change. *Environment and Urbanization*, 21: 225–240.

Burton, I. 2004. "Climate Change and the Adaptation Deficit." Occassional Paper 1. Adaptation and impacts research group, Meterlogical Service of Canada, Environment Canada.

CIF. 2010. Climate Investment Fund: Strategic Programme for climate resilience Bangladesh. Accessed May 12, 2013. http://www.climateinvestmentfunds.org/cif/sites/climateinvestmentfunds.org/files/PPCR%205%20SPCR%20Bangladesh%20nov2010.pdf

CIF 2011. Climate Investment Funds - PPCR - Fact Sheet September 2011.

Dasgupta, S., M. Huq, Z. H. Khan, M. S. Masud, M. Ahmed, N. Mukherjee, and K. Pandey. 2011. "Climate Proofing Infrastructure in Bangladesh: the Incremental Cost of Limiting Future Inland Monsoon Flood Damage." Journal of Environment and Development 20: 167–190.

Davidson, O., K. Halsnes, S. Huq, M. Kok, B. Metz, Y. Sokona, and J. Verhagen. 2003. "The Development and Climate Nexus: The Case of sub-Saharan Africa." *Climate Policy 3* Special Supplement on Climate Change and Sustainable Development.

GEF. 2012. "Bangladesh and GEF." Accessed October 10, 2013. www.thegef.org/gef/sites/thegef.org/files/publication/Bangladesh%20-%20Fact%20Sheet%20ready-FINAL.pdf

Government of Bangladesh (GoB). 2009a. Bangladesh National Adaptation Program of Action Plan. Dhaka: Ministry of Environment and Forests, Government of Bangladesh.

GoB. 2009b. *Bangladesh Climate Change Strategy and Action Plan*. Dhaka: Ministry of Environment and Forests, Government of Bangladesh.

GoB. 2009c. Policy Study on the Probable Impact of Climate Change on Poverty and Economic Growth and Options of Coping with Adverse Effects of Climate Change in Bangladesh. Dhaka: Planning Commission, Government of Bangladesh and UNDP Bangladesh.

GoB. 2014. "Bangladesh Climate Change Resilience Fund Project Sheet." Accessed October 10, 2013. http://bccrf-bd.org/Project.html

Haque, A. K. 2009. An assessment of Climate Change on the Annual Development Plan (ADP) of Bangladesh 15–11–09. Bangladesh: United International University.

Haque, U., M. Hashizume, K. N. Kolivras, H. J. Overgaard, B. Das, and T. Yamamoto. 2012. "Reduced Death Rates From Cyclones in Bangladesh: What More Needs to be Done?" *Bulletin of the World Health Organization*, 90 (2), 150–156.

Hedger, M. 2011. Climate Finance in Bangladesh: Lessons for Development Cooperation and Climate Finance at National Level. UK: Institute of Development Studies.

Huq, S., and J. Ayers. 2007. Critical List: The 100 Nations Most Vulnerable to Climate Change. Sustainable Development Opinion. London: International Institute for Environment and Development.

ICAI. 2011. The Department for International Development's Climate Change Programme in Bangladesh, Independent Comission for Aid Impact.

Khan, S. M. M. H., S. Huq, and M. Shamsuddoha. 2012. "The Bangladesh National Climate Funds. A brief history and description of the Bangladesh Climate Change Trust Fund and the Bangladesh Climate Change Resilience Fund." LDC paper series.

Lateef, F. 2009. "Cyclone Nargis and Myanmar: A Wake up Call."

MDMR. 2013. Assessment of Stakeholder's Role in Preperation for and Facing the tropical Mahasen, Comprehensive Disaster Management Programme. Dhaka: Ministry of Disaster Management and Relief, Government of Bangladesh.

MoEF. 2010. Climate Change Trust Fund Act 2010. Dhaka: Ministry of Environment and Forests, Government of Bangladesh. Accessed October 10, 2013. www.moef.gov.bd/Climate%20Change%20Unit/Climate%20Change%20Trust%20Act_2010.pdf

MoEF. 2012. Rio + 20: National Report on Sustainable Development. Dhaka: Ministry of Environment and Forest, Government of Bangladesh.

MoEF. 2013a. Approved project list of Climate Change Trust Fund. Ministry of Environment and Forests, Government of the People's Republic of Bangladesh. Accessed October 10, 2013. http://www.moef.gov.bd/Climate%20Change%20Unit/Approved%20Project%20-Update%20Up%20to%20April2013.pdf

MoEF. 2013b. BCCRF annual report 2012. Accessed September 1, 2013. http://bccrf-bd.org//Documents/pdf/BCCRF%20AR%202012%20-%20Final%20version%20-%206June13.pdf

MOF (Ministry of Finance). 2013. Bangladesh Economic Review-Bangla (2013). Economic advisor's wing, Finance division, Ministry of finance, Government of the People's Republic of Bangladesh. Accessed October 10, 2013. http://www.mof.gov.bd/en/index.php?option=com_content&view=article&id=249&Itemid=1

MoP. 2012. Public expenditure for climate change. Bangladesh Climate Public Expenditure and Institutional Review. Dhaka: General Economic Division. Planning Commission, Ministry of Planning.

Parry, M. L., O. F. Canziani, J. P. Palutikof, P. J. Van der Linden, and C. E. Hanson. 2007. "Climate Change 2007: Impacts, Adaptation and Vulnerability." Contribution of Working Group II to the Fourth Assessment Report of the Intergovernmental Panel on Climate Change. Cambridge: Cambridge University Press.

Paul, B. 2009. "Why Relatively Fewer People Died? The Case of Bangladesh's Cyclone Sidr." NatHazards 50: 289–304.

Pervin, M. 2013. "Mainstreaming Climate Change Resilience into Development Planning in Bangladesh: Country Report 2013."

Pervin, M., S. Sultana, A. P. I. F. Camara, V. M. Nzau, V. Phonnasane, N. Kaur, and S. Anderson. 2013. "A Framework for Mainstreaming Climate Resilience into Development Planning." IIED Working Paper.

Schipper, E. L. F. 2006. "Conceptual History of Adaptation in the UNFCCC Process." RECIEL 15 (1): 82–92.

SDC (Swiss Agency for Development and Cooperation). 2010. Disaster Risk Reduction Programme for Bangladesh 2010–2012. Bern: Swiss Agency for Development and Cooperation.

Transparency International Bangladesh (TIB). 2013. An Assessment of Climate Finance Governance: Bangladesh. Dhaka: Transparency International Bangladesh.

TIB. 2014. Challenges in Climate Finance Governance and the Way Out. Dhaka: Transparency International Bangladesh.

Managing rural landscapes in the context of a changing climate

Andrea Kutter and Leon Dwight Westby

Global competition for natural resources is intense and the supply of those resources is increasingly more constrained by climate variability and change. Governments and international development agencies have the dual responsibility to meet the socio-economic needs of the poorest and most vulnerable people while preserving and enhancing their natural capital. These responsibilities often are at odds with each other and different stakeholder groups have prioritised one over the other. This paper suggests that the landscape approach provides a solution for stakeholders to achieve climate change mitigation, adaptation, and poverty reduction goals, though not without some trade-offs.

La concurrence mondiale pour accéder aux ressources naturelles est intense et la fourniture de ces ressources est de plus en plus soumise aux contraintes de la variabilité et de l'évolution du climat. Les gouvernements et les agences internationales de développement ont la double responsabilité de pourvoir aux besoins socio-économiques des personnes les plus pauvres et les plus vulnérables, tout en préservant et en améliorant leur capital naturel. Ces responsabilités sont souvent incompatibles entre elles et différents groupes de parties prenantes en ont priorisé certaines plutôt que d'autres. Cet article suggère que l'approche du paysage donne une solution permettant aux parties prenantes d'atteindre les objectifs en matière d'atténuation des effets du changement climatique, d'adaptation à ce dernier et de réduction de la pauvreté, non sans quelques concessions toutefois.

La competencia por los recursos naturales a nivel mundial es intensa y la disponibilidad de dichos recursos se encuentra cada vez más limitada debido a la variabilidad y al cambio del clima. En este sentido, los gobiernos y las agencias de desarrollo internacionales tienen la doble responsabilidad de hacer frente a las necesidades socioeconómicas de las personas más pobres y más vulnerables, a la vez que deben conservar y elevar su capital natural. A menudo, ambas responsabilidades entran en conflicto entre sí y los distintos grupos de actores terminan haciendo prevalecer una sobre la otra. El presente artículo sostiene que el enfoque de paisajes puede convertirse en una solución para los actores, en el sentido de que pueden lograrse objetivos encaminados a mitigar el cambio climático, a fomentar la adaptación y a reducir la pobreza, no sin que deban existir algunas concesiones de ambas partes.

Introduction

We live in a world of increased competition for natural resources. In countries around the globe scarcities related to food, water, energy, and forests constrain social and economic opportunities with amplified severity. In this context can we still talk about win-win situations when it comes to satisfying all interests and needs equally? Ask one set of stakeholders what the most pressing issue is and what should be done to meet that need, and they might say that the key issue is producing more food with a view to empowering local communities. Ask another group and they might adamantly claim the most important issue is conserving the integrity of ecosystems as it provides the basis for people's livelihoods and the planet's health, including the preservation of biodiversity. Ask a third group and they will affirm that mitigating the root causes and impacts of climate variability and change should be at the centre of any debate, as the threat to the global climate translates into a threat to humankind.

The basic thesis of this article is that while stakeholders might agree on a common goal, like ensuring food security, ecosystem preservation, and addressing climate change, it is increasingly difficult to find a shared strategy unless trade-offs and compromises are made and accepted as a necessary part of the response. Further, the article strongly suggests that a jointly-owned consultation and dialogue platform to discuss priorities and agree on a strategy needs to be in place regarding how to move the shared agenda forward and minimise the acknowledged trade-offs.

The first part of the article offers three snapshots of how resource constraints are currently playing out worldwide. These short presentations provide the rationale for the proposed application of the landscape approach as the appropriate strategy to achieve a sustainable livelihood base for millions of people while conserving the health of our planet.

The second part briefly describes the sustainable livelihood concept and the landscape approach. The third section introduces the Climate Investment Funds (CIF), a response by the international community to the threat of climate change in absence of a global deal. We provide details on two CIF programmes which specifically address natural resource constraints and competition: the Forest Investment Program (FIP) and the Pilot Program for Climate Resilience (PPCR). The fourth part provides two practical examples from ongoing investments financed by the CIF. One example discusses the application of the landscape approach in the *Cerrado* biome in Brazil supported by FIP. The second example provides insights to an island-wide application of the landscape approach in St Vincent and the Grenadines, a small-island development state in the Caribbean. This intervention is supported by the PPCR.

The article concludes with some recommendations and open questions which hopefully will further the debate on the validity of the landscape approach as an appropriate strategy for achieving sustainable livelihoods in a resources-constrained world.

Setting the scene: the reality of natural resource constraints

The following three snapshots exemplify the impact of an unmanaged competition for natural resources exacerbated by climate change. These snapshots are short summaries of recent events and generalised to highlight the issue of concern:

- Snapshot 1: Achieving food security to ensure people's livelihoods
- Snapshot 2: Conserving biodiversity to maintain ecosystem balance
- Snapshot 3: Preserving forests to mitigate climate change

Snapshot 1: achieving food security

By 2050, nine and a half billion people worldwide will need to be fed. In practice, this means that agricultural production must increase by 60% per year (FAO 2012). Yet available land and water resources are finite and already under considerable stress. Climate variability and change has already had a dramatic impact on the agricultural cycle in many vulnerable countries of the world, leading to higher food prices and security volatility. This is especially true for farmers in low-income countries; challenged by low institutional and human capacities, and poor physical and communication infrastructure, they are often ill-prepared for the new reality of a changing climate (Godfray et al. 2010). In most cases, climate data and hydro-meteorological services are either not reliable or simply not available. Often, local communities and agri-businesses have no communication platform available to receive information on imminent or longer term climate-related stress such as floods, droughts, and fires. The challenges experienced by local communities to cope with the impacts of climate variability and change leads in many cases to a total surrender and migration to urban areas where the poverty cycle continues (Barrios, Berti-nelli, and Strobl 2006).

Snapshot 2: biodiversity conservation

Globally, an increasing number of plant and animal species are either endangered or facing extinction because of degradation or loss of critical habitat (Kuussaari et al. 2009; Butchart et al. 2010). Biodiversity is a key to ecosystem health. Human activity is often linked to the destabilisation of ecosystems and even the destruction of entire habitats, leading to a partial or total collapse of ecosystems. In many countries, the need for increasing food production has resulted in the expansion of area under agricultural land use, thereby encroaching into eco-systems such as forests or wetlands. This in turn has caused a significant decline in biodiversity and quality of other ecosystem services (McKee et al. 2004). Climate change has exacerbated the trend as species are not able to migrate in search of better habitat conditions, for example those in forest fragments located in larger agricultural landscapes (Opdam and Wascher 2004). This ultimately leads to their extinction. In other cases, people's need for food or for creating new or additional sources of income has actually caused the extinction or endanger-ment of many species through over-harvesting, including poaching and selective harvesting of high-value species.

Snapshot 3: preserving forests

Burning forests to clear land for agriculture, which becomes the new foundation for food or biomass production, releases huge amounts of greenhouse gases (Fearnside 2000). The role of forest ecosystems in stabilising the global climate and providing other important ecosystem ser-vices for people's well-being is well known. The destruction and degradation of all types of forests only exacerbates climate change and ultimately endangers the very base of people's liveli-hoods in the mid- to long term. It is often pointed out that preserving the remaining forests needs to be a priority, but examples of approaches to scale up effective solutions to protect forest eco-systems remain rare.

Taken individually, the agendas of organisations finding solutions for each situation described above are driven by passion and conviction for trying to reverse negative trends, all centring in one way or the other on securing people's livelihoods and the planet's health. When considering all three goals at the same time, however, a win-win strategy seems to be distant. It might be more appropriate to discuss a new paradigm for dialogue which is based on a discussion over trade-offs and finding a consensus on one strategy towards a shared common goal that incorporates elements of individual interests to the greatest extent possible.

The concept of sustainable livelihoods and the landscape approach

The concept of managing trade-offs in natural resources use is actually not new. The first time the idea of integrated land use planning and management was brought into the global political dialogue was in the Brundtland Report (WCED 1987). The idea was further promoted in 1992 with the adoption of the *Agenda 21*, a non-binding, voluntarily implemented action plan of the United Nations with regard to sustainable development (UNCED 1992). Ever since, principles and elements of that approach have been reiterated in countless fora and piloted at a small scale all over the world. However, even in the context of the rapidly changing global climate, sectoral fragmentation in planning, resources allocation, and policy-making is still a painful reality, and the scaling up of good practices and wide-scale adoption of lessons learnt on integrated natural resources management remains elusive. The landscape approach, which puts sustainable livelihoods at the centre of action, provides the strategic framework to support integrated natural resources planning and management.

The concept of sustainable livelihoods is based on the approach of coping with and recovering from stress, maintaining or enhancing natural assets, and providing sustainable opportunities for the next generation. The concept contributes to multiple net benefits at the local and global levels in the short and long term. It has become a paradigm for the survival of humankind in the context of a changing climate and increased competition for managing natural resources for short-term economic gain.

The wider landscape approach looks across large, connected geographic areas to understand natural resource conditions and trends, natural and human influences, and opportunities for resource conservation, restoration, and development. It seeks to identify important ecological values and patterns of environmental change that may not be evident when managing smaller, local land areas. Applying this approach, diverse public and private partnerships create sustainable and liveable communities that protect historic, cultural, and environmental resources. In addition, policymakers, regulators, and developers are encouraged to support sustainable land use planning and land management techniques that reduce unnecessary negative trade-offs and create a proactive and synergetic relationship between communities and natural systems.

The landscape approach follows the principles for linking livelihoods with adaptation measures and mitigation activities (Robledo, Kanninen, and Pedroni 2005). It encourages the:

(1) prioritisation of mitigation activities that help to reduce pressure on the natural resources;
(2) inclusion of vulnerability to climate variability and change as one of the risks to be analysed in mitigation activities;
(3) prioritisation of mitigation activities that enhance local adaptive capacity; and
(4) expansion of sustainability of livelihoods, with particular consideration for the poor.

Experience has shown that there are as many opportunities as there are challenges associated with the implementation of the landscape approach at the local level as a valid concept to secure people's livelihoods in a sustainable long-term oriented way. However, there is little evidence of the validity of the concept at scale yet, so that the development community still has to wait for proof of the effectiveness of the landscape approach to address global climate change mitigation and adaptation challenges in the context of sustainable development.

Testing the validity of the landscape approach at scale, the Climate Investment Funds (CIF) have been given the mandate to initiate transformational changes in economic sectors affecting or being affected by climate variability and change. The main goal of the CIF is to shift the trajectories of countries' economies towards inclusive and sustainable low-carbon growth. In order to achieve impact at scale, a programmatic approach has been deployed based on strategic partnerships and collaboration and, hence, the use of comparative advantages of diverse stake-holder groups in support of an agreed common vision. The CIF experience with the landscape approach as exemplified further below will provide some of the first insights and lessons on how to effectively employ this strategic framework with a view to address the challenges associated with climate change. These lessons will have potential to generate further evidence on the validity of the approach in a global context.

The Climate Investment Funds (CIF)

In 2008, the international community created the CIF, which were designed, on an interim basis, to scale up both climate financing and knowledge to support the transformation of economies that will equip poorer communities for a climate-stressed world and encourage emerging economies to rethink their development approach to present and future economic growth. Four programmes were designed in support of the agenda: the Clean Technology Fund, the Pilot Program for Climate Resilience (PPCR), the Forest Investment Program (FIP), and the Scaling up of Renewable Energy Program in Low Income Countries. Two of these programmes, FIP and PPCR, deal specifically with natural resources management across wider rural landscapes and, therefore, with the short- and long-term impacts of a changing climate on people dependent on the health of ecosystem services for their livelihoods and well-being.

Both programmes, FIP and PPCR, advocate the landscape approach as the underlying strategy for managing competing interests in support of sustaining people's livelihoods and improving their well-being, and addressing the global challenges associated with climate variability and change. Therefore these programmes, endowed with a combined budget about of about US$2 billion, will provide evidence on the validity of the landscape approach and collect lessons on how it responds to and addresses competing interests in a resource-constrained world.

The Pilot Program for Climate Resilience (PPCR)

The objective of the PPCR is to support countries' efforts to integrate climate risk and resilience into core development planning and implementation. The centrepiece for the PPCR is cross-sectoral planning and management of natural resources to ensure that vulnerable people can engage under rapidly changing conditions.

PPCR-supported investment programmes are country-led and build on National Adaptation Programs of Action (NAPAs) and other national development programmes and plans.

The PPCR supports nine countries and two regional efforts, which include some nine small-island development states.[1] US$1.3 billion has been allocated to these efforts. These pilot programmes have endorsed investment plans which provide a priority-based investment framework agreed by various stakeholder groups at the country level. Scaled up PPCR resources are deployed for the identified priority activities co-financed by other partners.

The Forest Investment Program (FIP)

The FIP supports developing country efforts to reduce deforestation and forest degradation and promote sustainable forest management that leads to emissions reductions and enhancement of forest carbon stocks (REDD+).

The FIP supports eight countries[2] in their efforts to address the drivers of deforestation and forestation. Forest types range from tropical moist forests to dry forest landscapes, including agro-silvo-pastoral systems which provide the livelihood base for millions of forest-dependent communities. US$640 million has been pledged to the FIP.

All eight countries have an agreed national vision for using large-scale FIP resources in the context of their national REDD+ priorities. Projects and programmes are being prepared and implemented in support of that common vision. Consultations at the country level remain the centrepiece for reaching the agreement.

Practical examples from implementing the landscape approach in FIP and PPCR

The following section describes examples from investments supported under the FIP and the PPCR in support of the landscape approach. The examples provide evidence of embedding the landscape approach in the investments to address climate change and achieve multiple benefits by proactively managing trade-offs in the wider rural landscape: (1) FIP: Investing in the *Cerrado*, Brazil; and (2) PCR: Watershed management in St Vincent and the Grenadines.

FIP: Investing in the *Cerrado*, Brazil

The Government of Brazil decided to invest its allocated US$70 million in FIP resources in the *Cerrado*, a vast tropical savannah eco-region of Brazil, covering mainly the states of Goiás and Minas Gerais. The main habitat types of the *Cerrado* include forest savannah, wooded savannah, park savannah, savannah wetlands, and gallery forests.

Brazil asserts itself as an urban-industrial country with an economy partly anchored in the export of primary products, including agricultural commodities. Its 2010 national inventory of GHG emissions showed that 77% of emissions are caused by the Land Use Change and Forestry (LUCF) sector. It also showed that in 2005, land use change in the *Cerrado* contributed 22% of net emissions, and estimates that this contribution has increased relative to the Amazon since deforestation levels in the Amazon have fallen more steeply than in the *Cerrado*. This information lead to the conclusion that land use change in the *Cerrado* was the most urgent mitigation challenge (Government of Brazil 2010).

Agriculture development in the *Cerrado* was a major driver of that economic growth, clearly compounding the mitigation challenge. The question before Brazil was whether there was an opportunity to develop the land resources and increase food production in Brazil in a manner that it also addressed the challenges associated with climate change. The answer was to find an approach that ensures that the Amazon continues to be a high priority conservation area, while the *Cerrado* continues to be a high priority production area throughout the wider landscape. Further research shows that land management in the *Cerrado* needs to be improved to enhance sustainability and further reduce GHG emissions. Motivating land users in the *Cerrado* biome to switch to more sustainable practices would reduce emissions, maintain the *Cerrado* as the priority area for agricultural production, and also create opportunities for employment. This is in line with Brazil's national development priorities such as poverty reduction, reduction of socio-economic inequality, and combating hunger.

The decision by the Government of Brazil to invest FIP resources in the *Cerrado* was based on considerations of exploring the full benefits of the landscape approach in the context of the challenge described above (Government of Brazil 2012).

It was expected that investments in the extensive savannah ecosystems of the *Cerrado* would be a strategic use of funds for the following reasons:

(1) it provides an alternative supply of land for intensified agriculture thereby reducing pressure to convert the Amazonian forests;
(2) it contributes to the development of land better suited for agriculture as a sustainable form of use; and

(3) it provides new, and extends existing livelihood opportunities for forest dependent peoples and local communities in the *Cerrado*.

Each of these reasons is explained in greater detail below.

(1) Alternative supply of land – increasing availability

Brazil is one of the world's top producers of high-demand agricultural commodities such as soybean, sugarcane, coffee, tobacco, and cattle. The impact of the industry is such that large-scale agriculture has been cited as a major driver of deforestation in the Amazon region (FAO 2007).

The Brazilian government has been trying, with demonstrated success, to curb this trend by implementing its Action Plan for Protection and Control of Deforestation in the Legal Amazon (PPCDAm). The PPCDAm actively promotes the reduction of deforestation in the Amazon by focusing on land and territorial planning, monitoring, control, and sustainable production activities; it is updated periodically to reflect changes in the dynamics of deforestation in the Amazon and lessons learnt from successful activities.

Recent detailed analyses of publicly available satellite photos show that Brazil has reduced deforestation rates in the Amazon over the past five years to lower GHG emissions (UCS 2011). The strategy being employed in the Amazon is highly effective and is now being complemented by the Action Plan to Prevent and Control Deforestation in the *Cerrado* Biome (PPCerrado). The PPCerrado, launched in September 2010, intends to extend and improve monitoring and control activities by federal agencies with the aim of environmental regulation of rural properties, enhanced sustainable forest management and fire-fighting, land use planning for biodiversity conservation, and water resources protection.

Furthermore, the PPCerrado will encourage the sustainable use of natural resources; promote environmentally-sustainable economic activities and the maintenance of natural areas; and restore degraded land.

The PPCerrado is a vital component of Brazil's landscape approach to forest conservation. It takes advantage of the conservation gains made in the Amazon, while recognising that conservation priorities in the *Cerrado* have been neglected earlier. In addition, there are opportunities for further resource development, sustainable income generation, and livelihood opportunities. With the projections that the global population numbers will continue to rise, there is no expectation that demand for agricultural land will abate. It is estimated that 1.8 billion hectares of the potential crop land available globally is located in developing countries and half of the total is concentrated in just seven countries, Brazil among them. There is virtually no spare land available for expansion in South Asia, the Near East, and North Africa. This means that much of future high-demand agricultural commodities will continue to be sourced from countries like Brazil.

The increased demand for agricultural products also signals a real threat to the conservation gains made in the Amazon. These growing pressures primed the *Cerrado* for new REDD+ investments, such as those promoted under the FIP. FIP financing will directly contribute to the development objective for the *Cerrado* for purposes of sustainable agriculture and carbon conservation. The combined effect of enhanced agricultural opportunities and a carbon conservation focus in the *Cerrado* will be to increase the availability of an alternative supply of land for agriculture production, helping to curb any future conversion of forest ecosystems in the Amazon.

(2) Land use suitability and sustainability

Nutrients are held throughout the soil profile under forests, but are largely held in mineral form and not directly accessible for food production. Once trees are removed, the nutrients held deeper in the soil are gradually released, and can sustain agricultural production for a couple of years. However, as rainfall levels tend to be greater over forests, nutrients become highly susceptible to leaching when the canopy is removed. This makes it difficult to sustain high levels of production in cleared forest lands over prolonged periods. Leaching is a slower process in grasslands and tends to sustain better levels of production over longer periods of time. Furthermore, grasslands have in history proven to be effective for conversion to agriculture – examples include the prairies in New Mexico, USA; the Pampas in Argentina; Rupununi in Guyana; and the Veld in South Africa.

The savannahs of the *Cerrado* have been successfully converted to an agriculture powerhouse but it has required extensive mechanisation and intensive soil inputs which have not been most favourable to the ecological integrity of the biome. This necessitates investments in approaches and technologies for sustainable use of the *Cerrado* for agricultural production.

Brazil will use over US$30 million in FIP resources to provide legal, financial, and technical assistance necessary for owner/occupiers of private landholdings in 11 states of the *Cerrado* to meet their obligations under new environmental laws. Direct support to enhancing greater compliance with environmental conservation requirements translates to wider use of good practices such as intercropping, reduced mechanisation, fallow cropping, etc. An additional US$9 million is designated for the implementation of a system to monitor vegetation cover and land use change in the *Cerrado*, *Caatinga*, and *Pantanal* biomes and an early-warning system for the prevention of forest fires.

(3) Livelihood opportunities

Another FIP investment will use US$11 million in support of credit lines for farmers and land users looking to improve and extend their operations in a manner that is more environmentally sustainable. An additional US$17 million will be spent in collecting and collating biophysical, economic, and socio-environmental data from approximately 5000 sample points to help private sector investors and public sector resource managers make better informed decisions in their agriculture and forestry business operations. The combined effect of both these investment areas is to enhance livelihood opportunities in a manner that reinforces balancing human development goals with sustainable land use.

Emerging lessons from FIP for the practical application of the landscape approach

Early successes in the practical application of the landscape approach in developing investment plans for the eight FIP pilot countries may be attributed to the emphasis on a clear presentation of the theory of change before investments actually start. The theory of change for FIP encourages countries to identify what particular drivers of deforestation and forest degradation they need to address and how best to address them in the context of their national livelihood and development goals. Identifying priority areas of action to address these drivers encourages public debate on the strategy for the use of FIP and other resources, and allows for a validation of the results of the application of the landscape approach over time by involved stakeholder groups. Full stakeholder engagement ensures a high level of ownership of the proposed actions that continues to manifest in the implementation phase. Listed below, are emerging lessons derived from developing the forest investment plans in the eight FIP pilot countries, all of which have used the landscape approach as the underlying framework for the proposed actions.

The following key lessons for the development of REDD+ interventions based on the landscape approach can be highlighted: (1) there is no blueprint for applying the landscape approach – country circumstances differ and call for flexibility to suit the particular needs and resources constraints; (2) an enabling environment, including the existence of a public forum for debate and consultation, for promoting new and scalable approaches to forest landscape management is needed; and (3) making informed decisions on natural resources management depends on the availability and quality of data and information which reflect the dynamic in the wider landscape – it provides the ability to identify opportunities for production and needs for conservation activities. Each lesson is described below in more detail.

Creating a space for flexibility of application of the approach

Natural resource management across wider landscapes for multiple benefits requires flexibility in the application of the approach which often may have to be modified taking into account country circumstances. The FIP Design Document acknowledges this fact and pilots the application of the landscape approach at scale to generate understanding and learning of the links between the implementation of forest-related investments, policies, and capacity development measures (FIP Design Document 2009).

Figure 1 maps the eight FIP pilot countries against the REDD+ continuum (CIF 2013). In this concept, the core piece of the REDD+ focuses on reducing emissions from deforestation (RED). Further elaboration of this conceptual framework considers that degradation of intact forests can be an important emission source of carbon as well (REDD). This framework is enhanced by the idea that improved forest management in standing forests, and the enhancement of carbon stocks through re-stocking or reforestation, are integral to the conceptual framework (REDD+).

The eight FIP countries prioritised financing for new and enhanced approaches to address the multiplicity of drivers of deforestation and forest degradation by taking into account their country-specific challenges regarding REDD+. All countries view the FIP investments as a contribution to larger programmes aimed at reducing emissions from deforestation and forest degradation, or programmes aimed at improved forest and agriculture management. As such, each country focuses its efforts on a piece of the larger REDD+ challenge that they face, and each country approaches the REDD+ challenge in its own strategic way, based on national context and other ongoing efforts.

Table 1 delineates the elements of REDD+ that each country has prioritised in its investment plan. The pilot countries vary with respect to how they strategically position interventions and activities. Some countries focus on direct drivers of changes in the forest areas, while other countries focus on proximate drivers that may reside outside the forest but are important factors that drive deforestation and degradation. It is apparent that the diversity in natural capital and socio-economic conditions affecting resource use in countries will require adapting

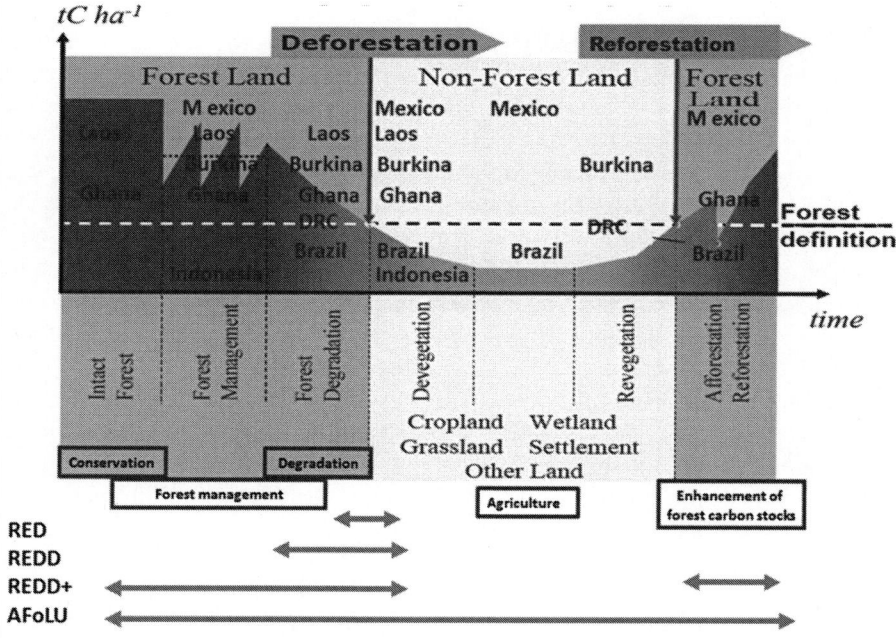

Figure 1.   Addressing the REDD+ continuum in FIP.

Table 1. FIP investment plans – elements of REDD+ and areas of investment.

Country	REDD+ elements addressed	Thematic scope of investments
Brazil	Deforestation	• sustainable agriculture • forest information systems • forest conservation • forest fire prevention
Burkina Faso	Sustainable forest management; deforestation; degradation	• sustainable forest management • fire management • community forestry • non-timber forest products
DRC	Sustainable forest management; degradation	• community forest management sustainable cook stoves • fuel-wood management
Ghana	Deforestation; degradation; sustainable forest management	• forest communities • sustainable forest management agroforestry (sustainable cocoa)
Indonesia	Deforestation; degradation; sustainable forest management	• forest governance • land tenure and indigenous rights law enforcement • sustainable forest management
Lao PDR	Deforestation; degradation; sustainable forest management and enhancement of forest carbon stocks	• sustainable forest management • community forestry • land tenure and rights of ethnic minorities • reforestation
Mexico	Deforestation; sustainable forest management; enhancement of forest carbon stocks	• community forestry • sustainable agriculture • rural development • sustainable forest management silvo-pastoral systems

the landscape approach to the specific country needs. A validation exercise with various stakeholder groups will verify that competing interests are addressed and negative trade-offs minimised. Brazil provides a good example of how the livelihood needs of rural communities, demands for agricultural products, and protection of forest reserves were integrated into the national policy to reduce emissions in the *Cerrado* while maintaining the biome as an agricultural powerhouse. This plan has wide support and will provide an excellent future opportunity to examine the effectiveness of the landscape approach.

Creating an enabling environment

Investing in institutional strengthening, governance reforms, and capacity building creates a sustainable foundation for the successful application of the landscape approach.

Improving governance and institutional capacities to address the challenges of deforestation and forest degradation is a top priority among many countries. To make the landscape approach a successful strategy to address REDD+, policy frameworks need to be improved, cross-sectoral dialogue fora need to be created at the national level, and laws and regulations need to be enforced. In all eight FIP pilot countries, inter-sectoral coordination and improving the legal and regulatory framework have been highlighted as essential to developing REDD+ engagement beyond the forest sector.

FIP pilot countries with low human and institutional capacity have opted to use FIP resources primarily for enhancing their REDD+ relevant policy frameworks and strengthening their ability to make informed decisions on their natural resources base and the drivers of deforestation and forest degradation. Building up the technical capacity to generate reliable data and information on land use dynamics in the wider landscape is a centrepiece for a successful application of the landscape approach to adress REDD+. This includes the ability of countries to monitor and verify the validity and implementation of policies and laws. Without an enabling environment, it would be difficult for stakeholders to have knowledge and access to tools and practices in support of the landscape approach. Lack of data and information would challenge the ability to monitor progress towards achieving the expected results and to learn what works and what does not.

Availability of data and information for informed decision-making

All FIP pilot countries have identified challenges when it comes to the availability of data for supporting a robust theory of change and the future evaluation of its validity. They acknowledge the need for better basic information and data on forest and land conditions, greenhouse gas data trends, and drivers of deforestation and degradation.

Another, often underestimated challenge related to data and information availability concerns the private sector. In Brazil for instance, a major barrier to private sector investment in sustainable forestry and land conservation management across the Amazonian landscape is the lack of information on resource diversity and availability and the potential market opportunities around them.

In an effort to remove this barrier, the FIP in Brazil will develop an integrated information system that will collect and collate this data and make it available to public and private sector agents. The project also purports to train these agents in the use and upkeep of the database and information systems to ensure the continued usefulness of the tool in the management of Brazil's forest and agricultural landscapes. Information is critical to the decision-making process where there is a complex set of sectors and actors competing for the natural capital of a country. For instance, information on the extent of forest resources, multiple-uses of those resources, the economic and other values attached to certain uses, help stakeholders to balance competing needs while managing forest ecosystems sustainably.

PPCR: St Vincent and the Grenadines

An illustrative case of the application of the landscape approach is a PPCR-funded intervention in St Vincent and the Grenadines. It is part of a regional programme designed for knowledge sharing and learning; for developing and applying standards and methods for climate change adaptation in watersheds, and for coastal protection in the Eastern Caribbean.

One part of the investment is implemented on Union Island as a trial experiment focusing a series of interventions on the island watersheds. The objective is to develop comprehensive climate-resilience interventions in small-island development states with a view to support sustainable livelihood options based on the sustainable management of the island's natural resources.

Comprehensive climate-resilience interventions respond to the multiple vulnerabilities of affected people. Similarly, as in the case of Brazil, explaining the theory of change underlying the investment design reveals a pattern that maybe used as the basis for investments in other islands. St Vincent and the Grenadines has embraced the opportunity to use the landscape approach as a strategy to satisfy both mitigation and adaptation objectives in the context of their sustainable development agenda.

As a first step, opportunities for natural resource development are identified that are responsive to the needs of the population. In the case of Union Island, this will be done by way of a full assessment of the island as a single drainage basin. In treating the island as an integrated landscape, from ridge to reef, for the first time, data and information on its natural resources and vulnerabilities as a whole will be available. The assessment will provide data on soil quality, ground water quality, downstream impacts of upstream activities, and shoreline stability.

Based on these results, supported by the PPCR, the application of Union Island's Integrated Coastal Zone Management plan and their community awareness strategy will be tested and assessed to ascertain how well the proposed interventions may enhance the adaptive capacity of the population.

The modified and updated Integrated Coastal Zone Management plan will contain a menu of interventions to be applied at the watershed level. These include the implementation of forest management activities, such as the replanting of mangroves and other plant species in selected areas. Other soil and water conservation measures will be promoted including the establishment of nurseries, terraces, and sedimentation traps; supporting good practices in agriculture and agro-forestry, and other activities.

The application of the landscape approach in the PPCR investment has the potential for multiple benefits: adaptation, mitigation, and livelihood benefits. Successfully maintaining the integrity of a watershed, its ecosystem base depends heavily on maintaining the vegetative cover since the water cycle relies on biological processes related to trees and plants. Hence, the conservation of biodiversity is another important element for success.

What can be learnt from the PPCR investment is that only when studying the vulnerability context of the watershed in the context of the whole island and understanding the various short-and long-term dynamics can an appropriate management plan be developed. This plan should provide opportunities for all land users while agreeing on trade-offs for the prioritised land use and management options. For instance, the vulnerability context of Union Island provides that agriculture and agro-forestry are both important economic sectors for the island and subject to a changing climate. Investments in hard infrastructure such as stone walls along the river banks or paved roads would not resolve the vulnerability of these sectors even though the impacts of physical threats, such as hurricanes and sea-level rise, would have been significantly reduced. Hence, the full vulnerability analysis for the island was the basis for designing investments in land management, soil and water conservation through mangrove planting, and supporting good practices in agriculture and agro-forestry with the ultimate goal to protect the livelihood base of the island population.

Lessons from PPCR for the practical application of landscape approaches

All PPCR pilot countries are vulnerable to the impacts of climate variability and change and the livelihood base of their population is at risk. PPCR countries, as in the case of FIP, have developed a jointly agreed investment framework to identify priority actions in support of their national development goals. Some countries have gone through extensive planning and consultation processes at the country level to ensure ownership for the proposed investments. These consultations were led by the government and attended by sector ministries, local communities, indigenous peoples groups, the business community, and development partners among others.

In all cases, the underlying strategy for the proposed PPCR-supported investments is the landscape approach as it has been recognised that climate variability and change has impacts on all economic sectors dealing with natural resources. Capacity development opportunities need to be offered so land users and other decision makers can understand the dynamics and be able to make informed decisions on land use and management in the context of a changing climate.

The example from St Vincent and the Grenadines shows the importance of identifying the opportunities for investment which respond to the multiple interests, needs, and vulnerabilities of those who are dependent on the use of natural resources. Once those dependencies and dependents are identified, resource management options can be discussed with them and agreed on through public dialogue. The final strategy arising from those discussions will be manifested in agreed land use or management plans and the communication of information on land use options which address the identified needs and vulnerabilities. In the case of St Vincent and the Grenadines, the strategy includes the design and building of drainage channels and buffer zones on the island, as well as engaging in a process of defining the legal and legislative implications for various communities. GIS mapping will be used to record the drainage systems. Such an undertaking in spatial planning will enable the implementation of the specific land use interventions and monitor its results.

Applying the landscape approach to natural resources management in climate-vulnerable areas, confirms and makes more compelling the potential this approach has in supporting the sustainable developing agenda.

Conclusions

The article discusses a world that is confronted with an increased competition for natural resources caused by development strategies focused on short-term economic gain coupled with rising renewable resource constraints. Scarcity of these resources is exacerbated by the impacts of climate variability and change. Few informed observers would question the fact that there is a need for more food, that delivery of ecosystems services must be stabilised and that the causes and impacts of climate change must be mitigated and managed. If these needs are not addressed, the livelihood base for billions of people will be further jeopardised.

This article has centred on the question of whether it is realistic to find a common strategy that enables all stakeholder groups to work towards a shared goal while, at the same time, accepting difficult trade-offs regarding each group's specific interests. We have responded that it is possible but only if stakeholders realise from the outset that difficult trade-offs will be a necessary part of finding a broad, societally accepted solution. We have recognised, in turn, that failure to fulfil all groups' vested interests can lead to social discontent and even instability.

The article proposes that the landscape approach offers a common strategy to provide for sustainable livelihoods on a short- and long-term trajectory integrating climate change considerations into the management of natural resources and people's behaviour. The landscape approach requires weighing various resource development options and understanding the full range of options and potential trade-offs. Investment decisions need to be discussed and agreed on among affected stakeholder groups with a common goal in mind. A common strategy needs to be agreed based on acceptable land management options and related trade-offs.

In fact, recent research by Lambin and Meyfroidt (2011, 1) affirms that this is indeed possible, and urges that "*land systems should be understood and modelled as open systems with large flows of goods, people, and capital that connect local land use with global-scale factors*".

The CIFs, through FIP and PPCR, have provided scaled-up resources to test whether the landscape approach provides a valid strategy to manage natural resources at scale in a sustainable way. Since implementation has only started, initial results pertaining to the effectiveness of the approach are still scarce. However, early experience suggests that applying the landscape approach at scale shows promise and provides lessons for how to best apply the approach in developing REDD+ investments. The CIF experience also invites a discussion on the following questions which should be posed when initiating the application of the landscape approach to sustainable natural resources management:

- How should a public dialogue and consultation on natural resource use at the country level be conducted and maintained?
- Who makes the decision as to which participants need to be included in the process? And who can best convene these dialogues?
- What are the roles and responsibilities of the participants in the consultation process?
- Do all stakeholder groups need to agree before a strategy for natural resources management can be implemented? How to deal with groups that disagree?
- What should be the structure and institutional setup for a redress/conflict resolution mechanism?
- What is the role of bi- and multilateral agencies in the consultation process?

It is our hope that as the implementation of FIP and PPCR investments further advance, more robust answers to these questions can be collected and documented. Through testing and direct application, we believe that the strengths of the landscape approach for managing landscapes at scale will become clear and persuasive.

Notes

1. PPCR pilots: Bangladesh, Bolivia, Cambodia, Mozambique, Nepal, Níger, Tajikistan, Yemen, Zambia. Caribbean regional programme (Dominica, Haiti, Jamaica, Grenada, St Lucia, and St Vincent and the Grenadines) and Pacific regional programme (Papua New Guinea, Samoa, and Tonga).
2. FIP pilot countries: Brazil, Burkina Faso, Democratic Republic of Congo, Ghana, Indonesia, Lao PDR, Mexico, Peru.

References

Barrios, S., L. Bertinelli, and E. Strobl. 2006. "Climatic Change and Rural-Urban Migration: The Case of Sub-Saharan Africa." *Journal of Urban Economics* 60 (2006): 357–371.

Brundtland, H., ed. 1987. *Our Common Future: Report of the World Commission on Environment and Development*. Geneva: the United Nations.

Butchart, S. H. M., M. Walpole, B. Collen, A. van Strien, J. P. W. Scharlemann, R. E. Almond, J. E. Baillie et al. 2010. "Global Biodiversity: Indicators of Recent Declines." *Science* 328 (5982): 1164–1168.

Climate Investment Funds. 2013. "Approaches to Measuring and Reporting Results in Endorsed FIP Investment Plans." Accessed October 21, 2013. https://www.climateinvestmentfunds.org/cif/sites/climateinvestmentfunds.org/files/FIP_SC.10_5_Overview_of_current_approaches_to_measuring_and_reporting_results_in_endorsed_FIP_investment_plans_0.pdf

FAO. 2007. "Sustainable Development and Challenging Deforestation in the Brazilian Amazon: the good, the bad and the ugly." Accessed October 21, 2013. http://www.fao.org/docrep/011/i0440e/i0440e03.htm

FAO. 2012. *World Agriculture Towards 2030/2050: The 2012 Revision*. ESAE Working Paper No. 12-03. Rome: FAO.

Fearnside, P. M. 2000. "Global Warming and Tropical Land-Use Change: Greenhouse Gas Emissions from Biomass Burning, Decomposition and Soils in Forest Conversion, Shifting Cultivation and Secondary Vegetation." *Climatic Change* 46 (1–2): 115–158.

Government of Brazil. 2010. *Second National Communication of Brazil to the United Nations Framework Convention on Climate Change*. Brasília: Ministério da Ciência, Tecnologia e Inovação (Government of Brazil).

Government of Brazil. 2012. *Forest Investment Plan for Brazil.* Brasília: Government of Brazil.

Godfray, H. C. J., J. R. Beddington, I. R. Crute, L. Haddad, D. Lawrence, J. F. Muir, J. Pretty, S. Robinson, S. M. Thomas, C. Toulmin. 2010. "Food Security: The Challenge of Feeding 9 Billion People." *Science* 327 (5967): 812–818.

Kuussaari, M., R. Bommarco, R. K. Heikkinen, A. Helm, J. Krauss, R. Lindborg, E. Ockinger et al. 2009. "Extinction Debt: A Challenge for Biodiversity Conservation." *Trends in Ecology and Evolution* 24 (10): 564–571.

Lambin, E., and P. Meyfroidt. 2011. "Global Land Use Change, Economic Globalization, and the Looming Land Scarcity." *Proceedings of the National Academy of Science of the United States of America* 108 (9): 3465–3472.

McKee, J. K., P. W. Sciulli, C. D. Fooce, T. A. Waite. 2004. "Forecasting Global Biodiversity Threats Associated with Human Population Growth." *Biological Conservation* 115 (1): 161–164.

Opdam, P., and D. Wascher. 2004. "Climate Change Meets Habitat Fragmentation: Linking Landscape and Biogegraphical Scale Levels in Research and Conservation." *Biological Conservation* 117 (3): 285–297.

Robledo, C., M. Kanninnen, and L. Pedroni, eds. 2005. *Tropical Forests and Adaptation to Climate Chnage: In Search of Synergies.* Bogor: CIFOR.

UCS. 2011. *Brazil's Success in Reducing Deforestation.* Briefing #8. Washington, DC: Union of Concerned Scientists (UCS).

UNCED. 1992. "Agenda 21: Programme of Action for Sustainable Development." Accessed October 21, 2013. http://sustainabledevelopment.un.org/index.php?page=view&nr=23&type=400&menu=35

Adaptation vs. development: basic services for building resilience

Fawad Khan

In a transect along Indus River after the 2010 floods in Pakistan, this article explores the relationship between the use and duration of use of basic services, among those who recovered well and those who did not, using non-parametric statistical testing in a quasi-experimental design. The research shows a clear and strong correlation between access and duration of usage of certain services before the disaster, and the rate of recovery in each location. This analysis demonstrates a relatively robust and cost-effective methodology to identify and prioritise development interventions that build resilience against climatic shocks that are not undertaken at the cost of poverty reduction.

Dans un transect le long du fleuve Indus après les inondations survenues en 2010 au Pakistan, les auteurs de cet article examinent la relation entre l'utilisation et la durée d'utilisation des services de base, parmi ceux qui se sont bien relevés et les autres, à l'aide de tests statistiques non paramétriques dans le cadre d'une conception quasi-expérimentale. Les recherches mettent en évidence une corrélation claire et forte entre l'accès et la durée d'utilisation de certains services avant la catastrophe, et le taux de relèvement dans chaque site. Cette analyse démontre une méthodologie relativement robuste et rentable pour identifier et prioriser les interventions de développement qui renforcent la résilience face aux chocs climatiques et qui ne sont pas entreprises aux dépens de la réduction de la pauvreté.

En 2010, en una parte del trayecto del río Indo en Pakistán se produjeron inundaciones. A partir del uso de pruebas estadísticas no paramétricas en un diseño cuasiexperimental, el presente artículo examina los vínculos existentes entre el uso y la duración de los servicios básicos en el sector de la población que se recuperó con facilidad y en aquel que experimentó dificultades. Al respecto, la investigación reveló la existencia de una fuerte y clara correlación entre el acceso a y la duración del uso de ciertos servicios previamente al desastre y la tasa de recuperación en cada localidad. El análisis utilizó una metodología relativamente robusta y rentable para identificar y priorizar las acciones de desarrollo que ayudan a construir la resiliencia ante los *shocks* climáticos, sin que éstas sean implementadas a costa de las actividades dirigidas a reducir la pobreza.

Introduction

The impacts of the 2010 Indus floods in Pakistan represent a fundamental challenge that crosses all aspects of life in the country. From livelihoods of rural populations to food supply to urban areas, systems such as transport, communication, energy, health, water control, and institutional systems on which populations depend, failed during the floods (Kronstadt, Sheikh, and Vaughn 2010).

While addressing the United Nation's General Assembly in connection to the floods, the Secretary General said: *"Almost 20 million people need shelter, food and emergency care. That is more than the entire population hit by the Indian Ocean tsunami, the Kashmir earthquake, Cyclone Nargis, and the earthquake in Haiti—combined"* (UN News Centre 2010). According to the joint damage needs assessment undertaken for the Government of Pakistan, the recovery was assessed to be in the order of US$8.74 billion to 10.85 billion (Asian Development Bank and World Bank 2010).

This paper is based on the findings of the project named "Building research capacity to understand and adapt to climate change in the Indus Basin, Pakistan" (Khan 2013). The International Development Research Centre (IDRC) and the Department for International Development (DFID) of the United Kingdom funded the project. It was implemented by the Institute for Social and Environmental Transition (ISET) and Rural Support Network Programme (RSPN) in Pakistan with its member rural support programmes (RSPs). Agha Khan Rural Support Programme (AKRSP) conducted the research in Chitral District, Sarhad Rural Support Programme (SRSP) in Charsadda District, and Thardeep Rural Development Program (TRDP) in Dadu and Tharparker Districts.

Research issue

Mounting evidence on climate change due to anthropogenic activities has raised concerns about the impact of these changes on poor and vulnerable populations (IPCC 2007). Along with international efforts for the mitigation of greenhouse gases, there is a growing concern over the impacts on developing nations who have contributed very little to the changes in the atmosphere. Apart from rapid advances in climate science, there has been a massive theoretical effort on developing the concepts of vulnerability, and lately, resilience. Within this debate, there have been three schools of thought that emanate from the risk/hazard literature, political economy/ecology perspective, and recently, from ecological resilience (Eakin and Leurs 2006, 369–371). Although all of these approaches are very rich in content and evolve rapidly, for a development policymaker and practitioner they pose a major problem in solving the key issues related to combating the negative impacts of climate change. These are summarised by Hinkel (2011, 198):

> "(i) identification of mitigation targets; (ii) identification of vulnerable people, communities, regions, etc.; (iii) raising awareness; (iv) allocation of adaptation funds; (v) monitoring of adaptation policy; and (vi) conducting scientific research. It is found that vulnerability indicators are only appropriate for addressing the second type of problem but only at local scales, when systems can be narrowly defined and inductive arguments can be built. For the other five types of problems, either vulnerability is not the adequate concept or vulnerability indicators are not the adequate methodology."

Moreover, the second type of problem may be identified at local scales but cannot be easily addressed. When vulnerability analysis identifies vulnerability through indicators such as gender, age, and social class, there is no indication to how this vulnerability is to be reduced. It tells us more about who is vulnerable as opposed to why they are vulnerable. Simply speaking, there is usually no diagnosis of the problem to be solved but rather an identification of a condition that can be irreversible and static.

A similar problem lies in the biophysical (climate) science side of climate change policymaking, where *"predict and prevent"* approaches are in great demand but of very little practical use in unpredictable climate-related disasters (Tyler and Moench 2012). Since we cannot accurately predict the time and intensity of the climate-based natural disasters, we cannot accurately and cost-effectively plan to prevent harm on these parameters.

This study uses the unfortunate event of 2010 floods in Pakistan to utilise the concept of resilience[1] as a basis for empirical observation for conducting research on the ability of people to recover from unpredictable and unforeseen climatic shocks. This is achieved through objectively measuring recovery and then statistically correlating it with use of basic services (treatments) as a driver of resilience that allows households and communities to build adaptive capacity and recover relatively faster, using a quasi-experimental approach.

Conceptual framework

We hypothesise that resilience is negatively affected where core gateway systems are fragile and subject to failure *and* where populations are socially or economically marginalised (ISET 2008; Ribot, Najam, and Watson 1996). Social and economic marginalisation is hypothesised as the primary factor limiting both access and the ability to use and obtain benefits from gateway systems. As a result, the impact, during flood or drought events, is likely to be highest where fragile systems and marginal populations overlap. This intersection makes those affected more vulnerable. Resilience, the reverse of vulnerability, will be highest where access to benefits that a population gets from systems and services is not constrained by social marginality.[2] This is illustrated in Figure 1.

In this framework, livelihoods of agents are supported by various systems, which include core systems such as water, land, and air; gateway systems such as drinking water, health, education, etc., and tertiary systems such as governance and institutions. These systems are further impacted by climate change. Using systems and services as a basis for analysis allows us to identify causes of vulnerability where it can be treated. Incorporation of core and gateway systems and their access as an integral part of vulnerability or resilience analysis allows us to identify critical ones that can be prioritised to define more resilient pathways for development.

Methodology

This section describes the mix method process adopted for the research. It describes the site selection process, sample size, the mix of qualitative and quantitative tools that were used, development of the recovery index, and hypothesis testing.

Our main hypothesis is that availability, use, and duration of use of services builds resilience (defined as the ability or capacity to recover following disruptive events such as the 2010 and 2011 floods) in communities. Specific hypotheses for testing were that:

(1) households with reliable access to basic systems and services will recover more quickly from the floods; and
(2) the duration for which services were used before the floods will increase their adaptive capacity and hence aid a relatively faster recovery.
(3) these basic "gateway" systems and services will differ among communities due to differences in geographic and socio-economic conditions.

We used a qualitative-quantitative-qualitative approach; the process followed to conduct the field study was:

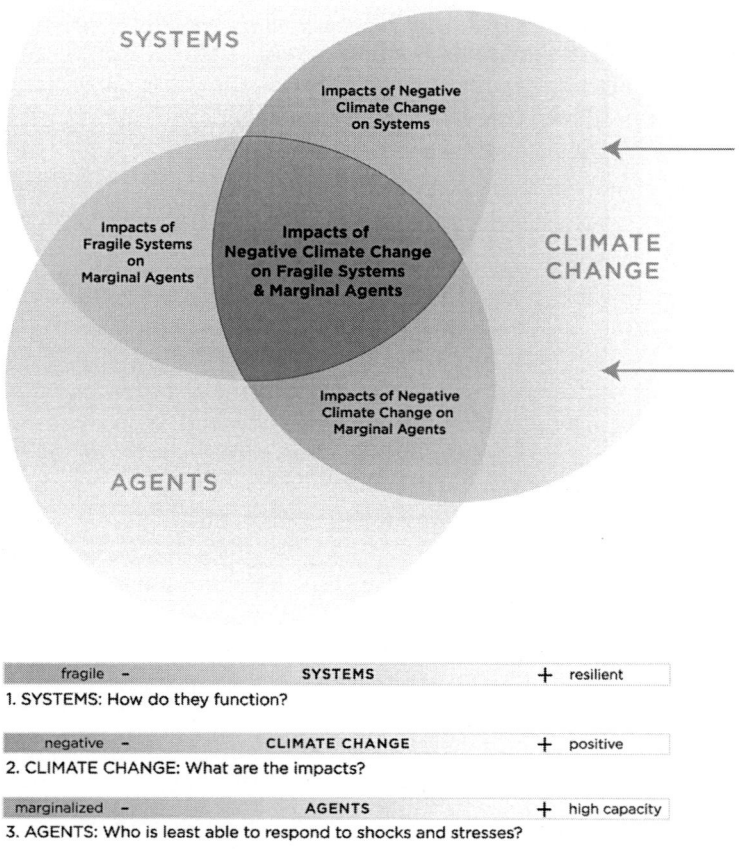

Figure 1. Systems, agents, and climate change.
Note: Adapted from ISET 2008.

(1) Identification of under-serviced areas/populations in flood-affected areas using census data to be able to empirically test the impact of those services between those who recovered well and those who did not.

(2) In post-disaster situations, identification of vulnerable and resilient populations through their recovery status using qualitative Shared Learning Dialogue process (Reed, Guibert, and Tyler 2011) with communities and triangulating it with quantitative asset recovery index.

(3) Quantitative documentation of service use (quantity, quality, time) with questionnaires at households selected through the process above.

(4) Analysis of quantitative data to identify differences in services that are statistically associated with rapid recovery and therefore can be interpreted as building resilience.

(5) Analysis of qualitative data to evaluate associations (or the lack thereof) between service usage and recovery. Evaluation of both quantitative and qualitative data to interpret potential relationships between services and resilience at the household and community level.

Site selection

To incorporate a large geographical variation along the Indus transect, the study sample includes the following areas (see Figure 2):

- High mountains (Chitral District)
- Indus-Kabul confluence and piedmont (Charsadda District)
- Plains (Dadu District)
- Desert/coastal (Tharparkar District).

These areas represent the major physical features of the Indus River basin. They include the upper reaches where glacial melt feeds the river, the piedmont, the plains, and the desert. In addition, they also cover a wide social and political spectrum in Pakistan and were severely affected by the 2010 floods, and 2011 floods in case of Tharparkar. Using such varied locations allowed us to examine whether the factors (services) leading to resilience were similar or varied in different geographical and socio-economic conditions.

These districts were categorised as "moderately affected" by UN-OCHA (2011) and offered areas where substantive damage was done to household stock, yet villages were not completely wiped out. At the same time, an index was created with the help of census data to identify two villages within each of the four selected districts that had relatively limited access and use of basic services before the floods to be able to test the correlation of service differential with recovery.

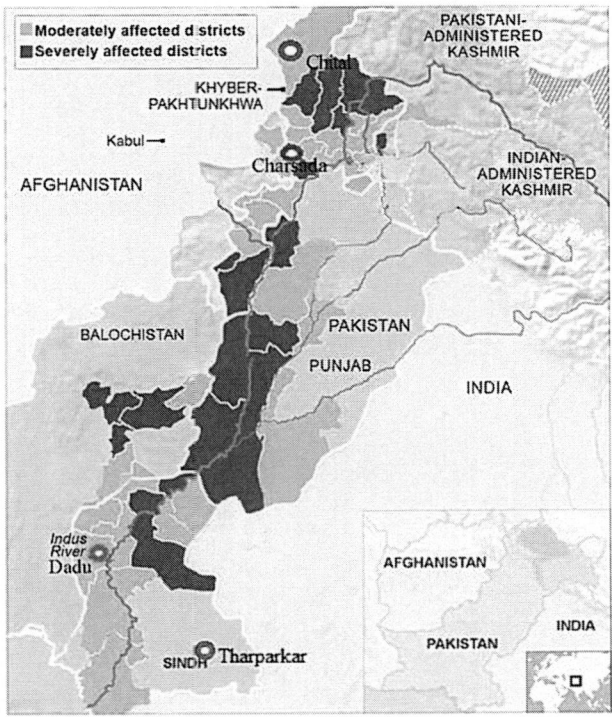

Figure 2.   Geographical location of selected districts.
Note: Adapted from UN-OCHA 2011.

Qualitative methods

The qualitative part of the study was conducted through a shared learning process (Reed, Guibert, and Tyler 2011) that employed participatory tools. The process and tools were jointly developed with field teams and Table 1 shows the type of data collected and the tools that were used. This exercise allowed us to document the local context in terms of geography and the livelihoods systems. Additionally, it highlights the climate-related hazards that are expected to impact livelihoods adversely.

In the shared learning process, the survey teams asked the community to jointly identify households (with equivalent exposure, i.e., level of flooding) that were recovering well and those that were relatively slow in recovery. In each village (location), at least 15 households were identified in each category. In some places, it was not always possible to have the requisite number of households so additional/neighbouring villages were included.

Table 1. Data collection tools.

Pre-steps (Revenue village level)	Comparison between most vulnerable and resilient groups (gender segregated)			
	Hazard	Livelihood	Availability and access to services	Social networks
• Identification, Selection of UCs and Villages	Ranking;	Diagrams: (livelihood Sources);	Listing of services/ timeline;	Narrative;
• Initial Contact– (Community Activists)	Responses;			Matrix of Institutional Mapping (organisation, community, network, linkages, etc.);
• Transect Walk	Proposed Solutions/ Ranking;	Calendars (Seasonal, agricultural, etc.);	Decision- Making Matrix;	
• System/Services				
• Vulnerability of services				
• Basic Information	Institutional options.	Daily Gender Workload Charts;	Mobility matrix (internal and external).	Political affiliation;
• Mapping				
• Natural Resources/ boundaries				
• Hazard Mapping		Decision Making Matrix.		
• Timelines (event/impacts)				
• Selection of households/ communities that have either the slowest or fastest recovery (with similar exposure to hazard)				

Data source: Khan 2013.

Quantitative methods

A total of 235 households were surveyed in all four districts using the purposive sampling criteria described above. Table 2 provides a summary of the basic characteristics of the sample used for this analysis.

Questionnaires

Three types of questionnaires were used in this research. A village level questionnaire was developed to record the basic facts about the villages, and more specifically, to record the availability of services in that village. More than 17 basic services including ecosystem services (such as land, forest, pastures), and basic gateway services (such as drinking water, education, health, and communication) were recorded. At the same time, the access and use of each service was recorded from each of the selected household through a separate questionnaire. The third household questionnaire recorded the housing assets before the flood, the assets immediately after the floods and current assets to develop the recovery index described below.

Recovery index

This index was created to measure the material recovery of the surveyed households. We used the damage data of housing as a measure for recovery in terms of how much of the housing assets were rebuilt using the following formula:

$$\text{Recovery rate (RR)} = \frac{\text{Increase or(recovery of assets) after the flood}}{\text{Loss of assets in the floods}}$$

Table 2. Summary profile of the sample households.

District/villages	Sample Households (no.)	Total Population	Household Size	Male: Female Ratio %	Adults per Household
Chitral District	60	415	6.9	105.4	4.9
Gouch	15	108	7.2	111.8	4.6
Madaklasht	15	85	5.7	80.9	4.4
Rambur	15	120	8.0	126.4	6.0
Sheikhandeh	15	102	6.8	100.0	4.7
Charsadda District	56	344	6.1	112.3	4.4
Agra	26	176	6.8	104.7	4.8
Kharkai	30	168	5.6	121.1	4.1
Dadu District	69	506	7.3	121.9	5.1
Luqman Shahani	28	206	7.4	114.6	5.3
Saeed Khan Shhahani	17	118	6.9	140.8	4.3
Seelaro	24	182	7.6	119.3	5.5
Tharparkar District	50	386	7.7	115.6	4.8
Bhakuo	30	214	7.1	137.8	4.8
Haryar	20	172	8.6	93.3	4.8
Grand Total	235	1651	7.0	114.1	4.8

Data source: Khan 2013.

$$RR(\%) = \frac{(Ac - Aa)}{(Ab - Aa)} \times 100\%$$

Where
  Aa = *Assets after the flood*
  Ab = *Assets before the flood*
  Ac = *Current Assets*

Among assets owned by families, housing structure is by far the most expensive and accounts for the majority of the material assets. Housing damage data is also readily available and easily verifiable, and therefore the housing structure was used as the proxy for material recovery. Moreover, since the research took place in a data-scarce environment, recall data had to be used, in which case housing structure damage and recovery was most likely to be remembered accurately. In Mithi District, where houses are made of mud, a locally available natural material, most houses had been reconstructed by the time the villages were surveyed. In that case, the recovery rate was measured by the date of reconstruction.[3] In all other sites, both quantity and quality of construction and the recovery was recorded and priced at unit cost in three types of construction, namely, *katcha*, *semi-pucca*, and *pucca*. Table 3 describes the recovery status of the surveyed households.

The recovery rate was used to triangulate the two groups identified by the community with material recovery. A number of households were eliminated from the sample on this basis. In the remaining sample, statistical testing was conducted to ensure that the extent of damage to the housing structure or the absolute size or quality of construction was not affecting the recovery rates.

Table 3. Household structure damage and recovery status in the sample households (HH) as of September 2011.

District/Village	Total HH surveyed	Number of HHs damaged	House structure damage recovery rate %				
			No recovery	1–50	51–75	75–100	>100
Chitral District	60	27	33.3	7.4	3.7	55.6	–
Gouch	15	8	50.0	12.5	–	37.5	–
Madaklasht	15	15	26.7	6.7	6.7	60.0	–
Rambur	15	4	25.0	–	–	75.0	–
Sheikhandeh	15	–	–	–	–	–	–
Charsadda District	56	51	27.5	21.6	–	49.0	2.0
Agra Payan	26	21	33.3	9.5	–	57.1	–
Kharakay	30	30	23.3	30.0	–	43.3	3.3
Dadu District	69	68	1.5	17.6	8.8	67.6	4.4
Luqman Shahani	28	28	3.6	14.3	7.1	64.3	10.7
Saeed Khan Shhahani	17	16	–	6.3	-	93.8	–
Seelaro	24	24	–	29.2	16.7	54.2	–
Tharparkar District	50	50	–	–	–	100.0	–
Bhakuo	30	30	–	–	–	100.0	–
Haryar	20	20	–	–	–	100.0	–
Grand Total	235	196	12.2	12.8	3.6	69.4	2.0

Data source: Khan 2013.

As the recovery index measures recovery in relation to the original asset base, it is not biased by the absolute value of the assets and controls for wealth. Poverty measurement and reduction is a topic that is of much importance and needs research on its own merit.

Hypothesis testing

As we had purposively sampled through the SLD process and then triangulated the households through the material recovery index, the sample sizes were small (between 5–25 for each group). Since these groups (the most and least recovered households) represented tails of the population distribution we had precluded the use of all tools that assumed a normal distribution. These sample characteristics led to the use of non-parametric testing to verify correlations. Chi-Square and Fischer exact tests were used to establish the difference in access to services in both groups at each site, with the caveat that they do not predict the direction of relationship.

Along with access to services, we also tested to see how the cumulative effect of having used a service over a period of time influenced the differentiation between the slow and fast recovered. The Man-Whittney U test was used to determine the statistical significance and direction of this correlation. These tests were performed for each service and its subsets. For example, it was used for communication but also for its subsets, which included telephone, cell phone, radios, televisions, etc.

Role of services in resilience

The following is a summary of results for selected services. These are discussed in more detail in Khan and Malik (2013) and also organised by location. This gives a much more nuanced picture of recovery and vulnerability and its contextual nature.

There is a clear statistically significant indication that when a few of the critical gateway services are missing, the communities' ability to recover from unexpected shocks is negatively impacted.

Electricity

Where there is no universal coverage, access to electricity was a major difference between those who recovered faster and those who did not. Electricity opens doors to many other services especially communications, etc., elongates the workday, and can have indirect effects like improving girls' enrolment in schools. Public sector professionals are also more likely to serve in areas with electricity and do not like to be posted in areas where electricity is not available.

Therefore, the availability of electricity increases the availability of services that in the long run help communities diversify livelihoods through skill enhancement, knowledge, and the ability to communicate, as we saw in Chitral (see Table 4). Here, people used seasonal migration as a strategy for reducing reliance on the forest. Cutting trees for income was making communities without external income even more vulnerable to floods and landslides. Supply of electricity had a similar effect in Tharparkar, which is a desert. Dadu and Charsadda District also had a higher number of well-recovered households with access to electricity, although we did not find the difference to be statistically significant in our sample.

Table 4.   Electricity.

| | Used before the flood | | | Duration of use before the flood | |
Service	% Using R: recovered; NR: not-recovered	Chi-Square p-value – 2 sided	Fisher's exact test probability of a chance occurrence – 2 sided	Average years used R: recovered; NR: not-recovered	Mann-Whitney U test p-value for (2 sided)
Chitral	R: 100%; NR: 40%	0.001***	0.001***	R: 16.00; NR: 4.47	0.001***
Bhakuo	R: 77%; NR: 57%	0.357	0.613	R: 18.00; NR: 0.00	0.022**

Data source: Khan 2013.
Note: ***represents a 99% level of significance; **represents a 95% level of significance.

Drinking water

Water supplies for drinking were categorised as piped water to house, hand-pumps, surface water, water courses/irrigation canal, and dug wells. Improved drinking water was a critical service in Dadu where people depend on poor quality surface water taken from irrigation watercourses. Households with access to hand-pumps and "sweet" ground water tended to recover faster. Such interventions could be useful for most of the Southern Punjab and Sindh where the poorest people in Pakistan live and where floods contaminate surface water supplies (Figure 3).

Figure 3 shows the use of hand pumps in two groups over time. The "recovered" group always had a larger proportion using improved water supply. Statistical testing for water is given in Table 5.

In Chitral, however, the piped water scheme did not seem to make a difference. Use of piped water was more recent among those who recovered faster. This result seemed spurious at first and needed further investigation. Chitral has drinking quality surface water in the form of numerous springs and snowmelt-fed streams (Kreditanstalt für Wiederaufbau 2006), so piped water did not improve the quality of this service, and therefore, did not a have an effect on the population's health and adaptive capacity. Forest plot of meta-analysis by Few-trell and Colford (2004) shows that water supply projects and reduction in diarrhoea do not always have a positive relation and can be problematic, especially if water is stored at

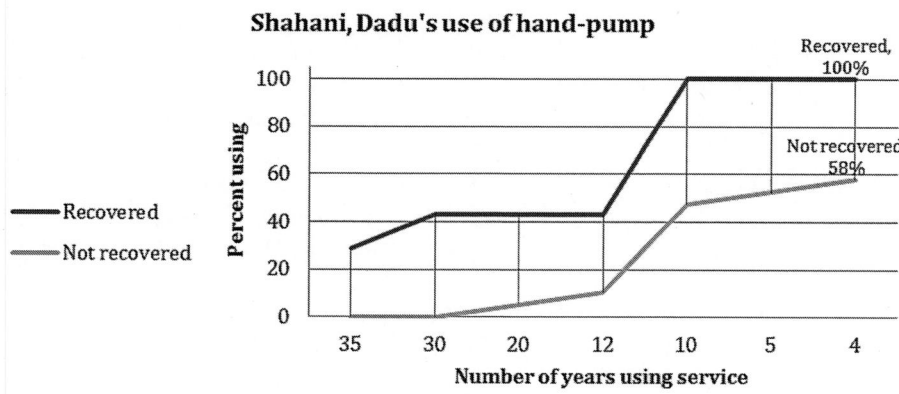

Figure 3.   Duration of hand-pump use in Shahani, Dadu.
Data source: Khan 2013

Table 5.   Drinking water.

Service	Used before the flood			Duration of use before the flood	
	% Using R: recovered; NR: not-recovered	Chi-Square p-value – 2 sided	Fisher's exact test probability of a chance occurrence – 2 sided	Average years used R: recovered; NR: not-recovered	Mann-Whitney U test p-value for (2 sided)
Chitral					
Piped to house (drinking)	R: 100%; NR: 100%		Not applicable	R: 6.00; NR: 19.00	$0.000^{(***)}$
Shahani					
Hand pump	R: 100%; NR: 58%	0.039**	0.062	R: 20.00; NR: 5.84	0.007***

Data source: Khan 2013.
Note: ( ) represents a negative correlation; ***represents a 99% level of significance; **represents a 95% level of significance.

home and/or there are poor hygiene practices. Use of latrines, however, did in some cases reduce the incidence of water-related diseases (Fewtrell and Colford 2004). In the next section, we see that the people in Madaklasht (recovered) had adopted the use of toilets much earlier than people in Gowch (less recovered). Therefore, improved water quality was seen to be an important factor in resilience and it was also linked to sanitation. In all other places, the water supplies were similar and we could not discern differences on the basis of quality and quantity.

Sanitation

Sanitation coverage is only 64% overall in Pakistan and 34% in rural areas (Water Supply and Sanitation Collaborative Council 2013). Sanitation services were broken down into drainage/ liquid disposal, solid waste removal, latrines/toilets, and street pavement. In Charsadda and Chitral, there was a clear difference in recovery rates related to duration of use. In Chitral, it was duration of use of latrines and in Agra, Charsadda, it was the difference in the duration of use of all of the four categories in sanitation (equally weighted aggregate). In the plains, however, drainage and street pavement seemed to be making the difference in the aggregate score. This does not reflect entirely on the efficacy of the intervention in building resilience but more on the fact that street pavement and drainage are more prevalent than other services (see Table 6).

Proper drainage along the streets is also important and where this did not exist, faeces was seen in the streets. Where there were drains and/or streets were paved, people (usually children) were less likely to defecate. Drainage is, of course, likely to be of great importance in densely populated areas where inundation flooding occurs and results in con-tamination of surface water sources. It is relatively less important in deserts where population densities are low and the potential for contamination of drinking water sources is also low (Figure 4).

In most locations, the coverage of latrines was too small for us to make statistically significant differentiation; however, the majority of households that used them had invariably fallen into the recovered category in Dadu and Charsadda. Therefore, sanitation seems to be the most common thread among those who recovered better from floods in all geographical areas.

Table 6.   Sanitation.

Service	Used before the flood			Duration of use before the flood	
	% Using R: recovered; NR: not-recovered	Chi-Square p-value – 2 sided	Fisher's exact test probability of a chance occurrence – 2 sided	Average years used R: recovered; NR: not-recovered	Mann-Whitney U test p-value for (2 sided)
Chitral Latrine/toilet	R: 100%; NR: 100%	Not applicable	R: 5.83; NR: 1.00	0.000***	
Agra Aggregate of drainage/water disposal; latrine/ toilet; solid waste disposal; street pavements	R: 48%; NR: 29%	0.098	0.146	R: 11.62; NR: 4.46	0.043**

Data source: Khan 2013.
Note: ***represents a 99% level of significance; **represents a 95% level of significance.

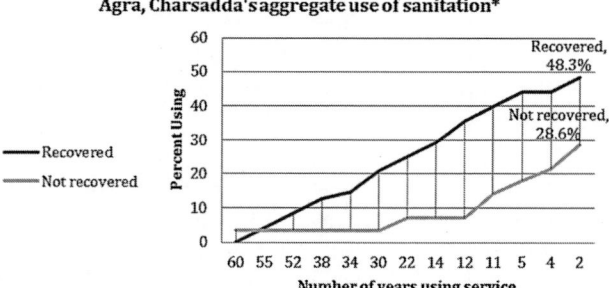

Figure 4.   Duration of latrine/sanitation use in Chitral and Charsadda.
Data source Khan 2013.
Note: *Aggregates drainage/water disposal; latrine/toilet; solid waste disposal; and street pavements with equal weights.

Health

Health was found to be important in all places and that corroborates the water and sanitation results (see Table 7). The quality of service seems to be particularly important. People who had access to better health facilities tended to recover faster. Lady health workers (LHW) seemed to be the healthcare providers of last resort and tended to those who could not access any other health facilities. Where better facilities outside the village were accessible through transportation, such as Kharkai and Bhakou, there was a negative relationship with LHW utility. In these places, roads are accessible and those with mobility (see section on mobility) can go elsewhere for healthcare. Those who relied on religious leaders also seemed to be worse off than those who did not.

Table 7. Health services.

Service	Used before the flood			Duration of use before the flood	
	% Using R: recovered; NR: not-recovered	Chi-Square *p*-value – 2 sided	Fisher's exact test probability of a chance occurrence – 2 sided	Average years used R: recovered; NR: not-recovered	Mann-Whitney *U* test *p*-value for (2 sided)
Chitral					
BHU/RHC	R: 0%; NR: 100%	0.000(***)	0.000***	R: 0.00; NR: 23.00	0.000(***)
Hospital	R: 75%; NR: 33%	0.031**	0.054	R: 13.75; NR: 9.87	0.243
Lady health worker/ health worker/ visitor	R: 83%; NR: 0%	0.000***	0.000***	R: 11.67; NR: 0.00	0.000***
Medical store	R: 75%; NR: 27%	0.013**	0.021**	R: 4.50; NR: 8.33	0.171
NGO clinic	R: 100%; NR: 0%	0.000***	0.000***	R: 20.00; NR: 0.00	0.000***
Religious leaders	R: 0%; NR: 100%	0.000(***)	0.000***	R: 0.00; NR: 50.00	0.000(***)
Veterinary clinic	R: 33%; NR: 0%	0.015**	0.028**	R: 5.67; NR: 0.00	0.018**
Kharkai					
Lady health worker/ health worker/ visitor	R: 64%; NR: 87%	0.159	0.215	R: 1.64; NR: 4.60	0.020(**)
Bhakuo					
BHU/RHC	R: 100%; NR: 100%	Not applicable	Not applicable	R: 19.85; NR: 19.00	0.021**
Lady health workers/ health worker/ visitor	R: 100%; NR: 100%	Not applicable	Not applicable	R: 8.46; NR: 9.57	0.038(**)

Data source: Khan 2013.
Note: ( ) represents a negative correlation; ***represents a 99% level of significance; **represents a 95% level of significance.

Given the health care coverage that visiting health workers provide, more emphasis on training and support of LHWs may help improve the provision of health services. The NGO clinics seemed to be providing better services than LHW and government's Basic/Rural Health Units in Chitral. This may be an alternative to the formal system for improving healthcare in Pakistan.

Credit and savings

Credit services played a key role in Charsadda, Chitral, and Dadu (see Table 8). In Chitral, it was used to finance the trip down country for seasonal migration and consumption smoothing over winter months. Similarly, people with access to credit had better recovery in both Dadu and Charsadda. In Charsadda, it was mainly credit from shopkeepers, whereas in Dadu it was formal credit from banks.

In Tharparkar, which has a very small cash-based economy and works more through barter, people reported that those who could access credit bought the scarce and expensive seed after the 2011 floods and were able to take advantage of the soil moisture and thereby recovered faster. Others had to wait till the following season, when seeds were more widely available. Although we could not statistically prove this relationship due to a small sample size seed availability had been a well-known factor in faster recovery. Di Falco, Veronesi, and Yesuf (2011), through a very complex experimental design, came to the conclusion that farmers in Ethopia with access to credit were much more likely to adapt by changing crops (through buying new seed) and benefitting from climate change.

This analysis shows how long-term credit in Chitral and short-term credit in Tharparkar was instrumental in recovery, but it is not an evaluation of microcredit services.

Table 8. Credit services.

Service	Used before the flood			Duration of use before the flood	
	% Using R: recovered; NR: not-recovered	Chi-Square p-value – 2 sided	Fisher's exact test probability of a chance occurrence – 2 sided	Average years used R: recovered; NR: not-recovered	Mann-Whitney U test p-value for (2 sided)
Chitral					
Banks	R: 25%; NR: 0%	0.040**	0.075	R: 1.17; NR: 0.00	0.044**
Community/ village organisation	R: 33%; NR: 0%	0.015**	0.028**	R: 1.17; NR: 0.00	0.018**
Shopkeepers	R: 69%; NR: 33%	0.058	0.128	R: 18.08; NR: 6.53	0.004***
Agra					
Shopkeeper	R: 58%; NR: 0%	0.011***	0.017***	R: 3.67; NR: 0.00	0.017**
Shahani					
Banks	R: 29%; NR: 5%	0.099	0.167	R: 1.71; NR: 0.00	0.017**

Data source: Khan 2013.
Note: ***represents a 99% level of significance; **represents a 95% level of significance.

Mobility and transport

In Chitral, once again, results of vehicle ownership (cars and 4 × 4s) were counterintuitive where both use and duration of use were negatively related to recovery (see Table 9). Residents of Gowch (less recovered) are transporters employed in the lumber industry and own comparatively more vehicles as opposed to the majority of Madaklasht residents (better recovered) who practise seasonal migration and have non-climate-dependent incomes. When floods strike the families of seasonal migrants, they do not lose any income whereas the owners of local transporters lose their mobility and income when roads are washed away for some time. Also, services such as schools, NGO clinics, and medical stores are available in Madaklasht and remain usable during and after floods. So in this case, transportation does not translate into better access.

In areas with good road networks, communities are able to access better services if they could access transport (see health results from Kharkai and Bhaku). This may be restricted by cost, as we saw in Charsadda, where only those who could afford it were using it to access better services. Social norms where female children are not allowed to attend schools that are outside the village restrict them from using transportation services to their advantage.

Education

The impact of education on recovery was analysed using four variables, namely, literacy (5+ years), literacy (18+ years), enrolment rate (i.e., percentage of school-going age children enrolled

Table 9. Mobility and transport.

Service	Used before the flood			Duration of use before the flood	
	% Using R: recovered; NR: not-recovered	Chi-Square p-value - 2 sided	Fisher's exact test probability of a chance occurrence – 2 sided	Average years used R: recovered; NR: not-recovered	Mann-Whitney U test p-value for (2 sided)
Chitral Mobility					
Access of main road	R: 100%; NR: 100%	Not applicable	Not applicable	R: 34.00; NR: 23.00	0.000***
Transportation Car	R: 0%; NR: 100%	0.000(***)	0.000***	R: 0.00; NR: 9.00	0.000(***)
4 × 4	R: 100%; NR: 100%	Not applicable	Not applicable	R: 24.00; NR: 29.00	0.000(***)
Public transportation	R: 100%; NR: 100%	Not applicable	Not applicable	R: 24.00; NR: 29.00	0.000(***)
Agra Mobility					
Access to main road	R: 100%; NR: 71%	0.050**	0.123	R: 36.92; NR: 17.86	0.042**
Transportation Animal	R: 42%; NR: 0%	0.047**	0.106	R: 13.92; NR: 0.00	0.097
Bicycle	R: 42%; NR: 0%	0.047**	0.106	R: 5.50; NR: 0.00	0.056

Data source: Khan 2013.
Note: ***represents a 99% level of significance; **represents a 95% level of significance; ( ) represents a negative correlation.

in schools), and average years of education. It is surprising to note that of all the locations, only two showed a significant statistical relationship in average years of education with recovery, and one of them was negative.

Comparing only literacy indicators and enrolment, we found a correlation in every district but the results were again more confounding than instructive (see Table 10). Strictly speaking, Chi-square tests and Fisher's exact tests cannot be interpreted for direction of relationship, though the percentages indicate a reverse relationship of recovery with literacy in both remote areas of Haryar (desert) and Chitral (mountainous). In contrast, the percentages in Agra and Kharkai indicated a positive relationship. Enrolment rate in Chitral was also significant and had the expected ratios in percentage.

Scrutiny of gender data revealed that the anomalies may have been caused by the large number of men that were unaccounted for in both remote locations (see Table 1). Due to the lack of local economic opportunities in these locations, they exhibit high rates of seasonal and career-long migration. Despite instructions to enumerators to record all people that contribute to the household economy, much like Sen's "missing women", the households consistently did not include these longer-term migrants, who are mostly male. The majority of these "missing men" were from the recovered households, the gender statistics revealed. It is possible that a lot of these long-term economic migrants may not have regular contribution to the livelihood but may have helped the recovery in kind, cash, or loans to their families. The households may also have not reported their contribution with the assumption that this may affect their eligibility for recovery support. Therefore, this phenomenon begs a more detailed empirical analysis before making conclusions.

These findings are somewhat mixed, as one would expect literacy and education to be key ingredients of developing adaptive behaviour and diversified livelihoods. Although we see a clearer relationship of literacy with recovery, the relationship of years of formal education and recovery is complex and contextual.

Table 10. Education.

Service	Literacy before the flood			Average years of education	
	% R: recovered; NR: not-recovered	Chi-Square p-value – 2 sided	Fisher's exact test probability of a chance occurrence – 2 sided	Average years of education R: recovered; NR: not-recovered	Mann-Whitney U test p-value for (2 sided)
Chitral				R: 2.63; NR: 3.12	0.143
Literacy 5+ yrs	R: 40; NR: 67	0.001***	0.001***	Not applicable	
Literacy 18+ yrs	R: 38; NR: 61	0.019**	0.030**	Not applicable	
Enrolment	R: 80%;NR: 55%	0.030**	0.048**	Not applicable	
Agra				R: 3.24; NR: 1.74	0.041**
Literacy 5+	R: 55%; NR: 27%	0.007***	0.012**	Not applicable	
Kharkai				R: 2.47; NR: 1.85	0.225
Enrolment	R: 57%; NR: 35%	0.038**	0.051	Not applicable	
Haryar				R: 2.84; NR: 4.53	0.017(**)
Literacy	R: 59%; NR: 78%	0.039***	0.057	Not applicable	

Data source: Khan 2013.
Note: ***represents a 99% level of significance; **represents a 95% level of significance; ( ) represents a negative correlation.

Caveats and nuances

Due to the site selection criteria, the results are biased to disasters that are intense enough to destroy damaged houses but do not completely destroy housing and communal assets, or those, which are minimally damaged.

As local people identify the households that recovered well and ones that did not, the samples from each location were biased by their perceptions of the recovery at that location. A quantitative recovery index was used to verify recovery, but it was based on material recovery only.

For services, statistical testing may under-represent their role in locations where there is either universal or very low level access of that service for lack of comparison. In the case of universal access, the continuation of provision may be an issue relating to resilience and where very few households have the service, statistical testing did not capture the difference. For example, only three households had title to their land in Shahani, Dadu, and all three had recovered much faster than others, but the sample size was not large enough to make a statistically significant inference.

Several basic development sectors have been analysed in this research. The purpose was to examine their potential relevance in building resilience to highlight a very pertinent aspect of adaptation that has not entered the academic discourse on vulnerability and adaptation. These results should not be viewed as an evaluation of these sectors or their relative performance as development interventions. Sectoral reviews need much more in-depth analysis and conducting them at the time of recovery from a historic disaster is neither a technically sound approach nor ethically appropriate.

In this study, we purposively sampled locations to ensure that there was no significant difference in wealth among surveyed groups of households in each location. This allowed for comparison of treatments (services) without the effect of wealth. So any interpretation of identification of priority sectors for adaptation in these locations as being equally effective in poverty reduction is not implied.

Finally, the recovery index may be a good proxy for material recovery but it does not cater to other types of non-monetary damages such as psychological impact that was reported at all locations. Trauma, loss of loved ones and physical disability caused by the floods was not captured in this analysis, and it constitutes a large part of human suffering in such large-scale disasters.

Conclusions

From the above analysis, it would be logical to assume that the particular services identified in different areas allow people to develop adaptive capacity in terms of income diversification and changing strategies as calamity hits them. A corollary to this hypothesis would be that those who do not have these basic gateway services before the disaster are slower to recover. For adaptation and development planning, this has significant implications. Since most climate-related hazards are unpredictable, most of the adaptation and recovery is expected to be autonomous; therefore, the vulnerable and exposed communities need to have basic adaptive capacity to recover. Despite the record UN appeal for recovery, we saw it had a very small role in the actual recovery of most households that we surveyed, and it was up to their own resilience as to how fast they recovered and adapted to the new realities after the floods.

The impact of a service on livelihood depends on its availability and access, duration of use, and, finally, quality. We were able to get much better results by assessing the duration for which the services were used. It was difficult to measure quality among all services, but where it could be compared, such as in healthcare and drinking water, the results were reliable and robust.

In rural agro-pastoral systems, we can also say that not all development interventions necessarily build resilience, and one has to be careful in identifying the critical interventions as many can also be maladaptive and exacerbate the pressures that populations and the climate are putting on our ecosystems. For example, in droughts, digging deeper and pumping out water may work in recovery in the short term but may actually increase vulnerability to future events (Khan and Mustafa 2008). All good adaptation strategies should reduce pressure on natural resources and possibly conserve the threatened resource.

The methods used in this study can be useful in prioritising critical gateway services for any disaster-struck area and offer a viable methodology for damage and recovery assessments for future development and resilience planning and financing. Since the method depends on lessons from disaster recovery, its use will be limited to areas that have recently been hit with disaster. However, we are also aware that disasters are catalysts for bringing investment into areas that are neglected in the usual development processes, and hence, this methodology can be instrumental in utilising disaster and adaptation-related funds to support much needed development of those areas and build resilience to future risks as a long term co-benefit.

Implications

The results of recovery from Indus floods in 2010 show that the complexity of the theoretical debate on definition of vulnerability has reached a level where most practitioners can scarcely elicit any practical lessons from the research. With increased funding for research on climate change, vulnerability has taken off as a field of its own with specialised qualifications and researchers. In this theoretical frenzy, the significance of understanding the livelihoods and basic development needs of the poor people is almost neglected. Since most of climate vulnerability research is theoretically driven in the North when the vulnerability also exists in the South, there is a need to solicit more empirical observations to base a theory on. Climate-related disasters affect millions of people every year, and so there is no dearth on the supply side of data and observation. Such evidence can help us find better ways to reconstruct and chose development options that are more strategic and cost-effective in terms of contributing to resilience of communities.

This research shows that all schools of thought in terms of risk/hazard literature, political economy/ecology perspective, and ecological resilience are theoretically relevant, but the concept of resilience has a particular advantage of being observable, measurable, and hence more suitable for practical application. We also see that development of a single theory driven, predictive vulnerability indicator or index is not possible, given the contextual complexity of livelihoods and unpredictable nature of climate related disasters. At the same time it would be unfair to suggest that this article has not benefitted immensely from the ideas proposed by the wide variety of vulnerability literature to formulate its hypothesis.

Critics of a resilience approach, especially engineering resilience, may argue that resilience in this sense would mean going back to a poor and disadvantaged state, at best. In reality, however, when the houses were rebuilt, without fail, the plinth levels were raised above the highest flood in memory. So people do adapt autonomously to newly perceived futures and build back better. Also, not falling into the poverty trap is an underachievement that is still to be preferred to the current conditions. This research also implies that indeed, development-based adaptation investments can improve ability to recover and contribute to poverty reduction rather than compete with it.

One interpretation of reverting to development interventions may be that we refocus on human development or the Millennium Development Goals. However, this research highlights that resilience needs are varied. The Human Development Index, developed by very educated humans, almost arbitrarily assigns equal weight for health and education. Whereas, in this case, the impact of health services showed a more direct impact both in terms of curative healthcare and in preventive measures such as water and sanitation. In contrast, the evidence of a relationship between education (as a measure of modern knowledge) and resilience was not nearly as clear as compared to that of literacy (which is more of a measure of communication). In another recent study in Pakistan based on a very large sample, Heltberg and Lund (2008) have also concluded that 55% of shocks suffered by poor families are health related. Their study shows:

> "high incidence and cost of shocks borne by households, with health and other idiosyncratic shocks dominating in frequency, costliness, and adversity. Sample households lack effective coping options and use mostly self-insurance and informal credit. Many shocks result in food insecurity, informal debts, child and bonded labour, and recovery is slow." (Heltberg and Lund 2008, 889)

In light of these findings, we can probably rethink some of the development priorities and the values attached to them. For the lack of more practical choices, resilience should become a part of the development objectives because vulnerability research has reached a point of sublimation and it needs to be grounded and guided by empirical evidence. Only then can it cater to the requirements of the vulnerable populations, which need to become resilient to the world that *they* live in.

Acknowledgements

We are deeply indebted to the recovering communities that offered their cooperation at a time of great loss in conducting this research, with the hope that it may pave the way for many other communities to be more resilient to the increased occurrence of climatic disasters.

Notes

1. That is, ability to recover according to the IPCC definition.
2. This is very similar to the argument Sen (1981) made in relation to floods and food security in Bangladesh.
3. The data showed that houses were rebuilt after harvests so they could be easily grouped by harvests after which they were built.

References

Asian Development Bank, and World Bank. 2010. Pakistan Floods 2010: Preliminary Damage and Needs Assessment. Pakistan: Asian Development Bank. Accessed July 10, 2013, http://reliefweb.int/report/pakistan/pakistan-floods-2010-preliminary-damage-and-needs-assessment

Di Falco, S., M. Veronesi, and M. Yesuf. 2011. "Does Adaptation to Climate Change Provide Food Security? A Micro-Perspective from Ethiopia." *American Journal of Agricultural Economics* 93 (3): 829–846. doi:10.1093/ajae/aar006.

Eakin, H., and A. Lynd Luers. 2006. "Assessing the Vulnerability of Social-Environmental Systems." *Annual Review of Environment and Resources* 2006 (31): 365–394. doi:10.1146/annurev.energy.30. 050504.144352.

Fewtrell, L., and J. M. Colford Jr. 2004. "Water, Sanitation and Hygiene: Interventions and Diarrhoea: A Systematic Review and Meta-Analysis." Health, Nutrition and Populaiton (Hnp) Discussion Paper. Washington, DC: The World Bank.

Heltberg, R., and N. Lund. 2008. "Shocks, Coping, and Outcomes for Pakistan's Poor: Health Risks Predominate." *Journal of Development Studies* 45 (6): 889–910.

Hinkel, J. 2011. "'Indicators of vulnerability and adaptive capacity': Towards a clarification fo the science-policy interface." *Global Environmental Change* 21: 198–208. doi:10.1016/j.gloenvcha.2010.08.002.

IPCC. 2007. *Climate Change 2007: Synthesis Report. Contribution of Working Groups I, II and III to the Fourth Assessment Report of the Intergovernmental Panel on Climate Change (IPCC Ed.).* Geneva: IPCC.

ISET. 2008. *From Research to Capacity, Policy and Action: Enabling Adaptation to Climate Change for Poor Populations in Asia through Research, Capacity Building and Innovation.* Kathmandu: Institute for Social and Environmental Transition (Nepal).

Khan, F. 2013. *Indus Floods Research Project: Final Technical Report.* Boulder, CO: Institute for Social and Environmental Transition (ISET).

Khan, F., and S. Malik. 2013. *Indus Floods Research Project: Results from the Field.* Boulder, CO: Institute for Social and Environmental Tranisition (ISET).

Khan, F., and D. Mustafa. 2008. *Final Report Desakota, Part II F4. Pakistan Case Study.* UK: NERC Science of the Environment.

Kreditanstalt für Wiederaufbau. 2006. "Drinking Water Supply and Sanitation Measures in Northern Upland/ Chitral District." Accessed October 25, 2013, www.kfw-entwicklungsbank.de/migration/ Entwicklungsbank-Startseite/Development-Finance/Evaluation/Results-and-Publications/PDF-Dokumente-L-P/Pakistan_Chitral_District_2006.pdf

Kronstadt, K. A., P. A. Sheikh, and B. Vaughn. 2010. *Flooding in Pakistan: Overview and Issues for Congress* (C. R. Service, Trans.). Washington, DC: UNT Digital Library.

Reed, S. O., G. Guibert, and S. Tyler. 2011. "The Shared Learning Dialogue: Building Stakeholder Capacity and Engagement for Climate Resilience Action." In *Catalyzing Urban Climate Resilience: Applying Resilience Concepts to Planning Practice in the ACCCRN Program*, edited by M. Moench, S. Tyler and J. Lage, 125–50. Boulder, CO: Institute for Social and Environmental Transition.

Ribot, J. C., A. Najam, and G. Watson. 1996. "Climate Variation, Vulnerability and Sustainable Development in the Semiarid Tropics." In *Climate Variability, Climate Change and Social Vulnerability in the Semi-arid Tropics*, edited by J. C. Ribot, A.R. Magalhães and Panagides. 13–51. Cambridge, UK: University of Cambridge Press.

Sen, Amartya. 1981. "Ingredients of famine analysis: availability and entitlements." *The Quarterly Journal of Economics*, 96(3): 433–464.

Tyler, S., and M. Moench. 2012. "A Framework for Urban Climate Resilience." *Climate and Development* 4 (4): 311–326. doi:10.1080/17565529.2012.745389.

UN News Centre. 2010. "Remarks to General Assembly Meeting on 'Strengthening of the Coordination of Humanitarian and Disaster Relief Assistance of the United Nations, including Special Economic Assistance'". UN Daily News. Accessed June 25, 2011. www.un.org/apps/news/infocus/sgspeeches/ statments_full.asp?statID=916 - .UtzstJGq6P_

UN-OCHA. 2011. "Updates: Relief Web". UN-OCHA Relief web. Accessed May 25, 2011. http://reliefweb. int/sites/reliefweb.int/files/resources/Full%20Report_737.pdf

Water Supply and Sanitation Collaborative Council. 2013. "Trends and Targets: Improved Sanitation and Drinking Water Coverage." Accessed October 2013. www.wsscc.org/countries/asia/pakistan/trends-and-targets

Supporting local climate adaptation planning and implementation through local governance and decentralised finance provision

Virinder Sharma, Victor Orindi, Ced Hesse, James Pattison, and Simon Anderson

Policies developed at national levels can be unresponsive to local needs. Often they do not provide the rural poor with access to the assets and services they need to allow them to innovate and adapt to the ways that increased climate variability and change exacerbate challenges to basic securities – food, water, energy, and well-being. In development deficit circumstances, common pool resources are important for climate adaptation purposes. In order for climate adaptation actions to deliver resilience, local perspectives and knowledge need to be recognised and given due priority in formal planning systems. Basing formal adaptive development planning on local strategies can support and strengthen measures that people have been tested and know to work. Local climate adaptation through collective action can address current increases in climate variability, future incremental changes, and the need to transform existing systems to deal with qualitative shifts in climate. These types of adaptation can work in cumulative ways. The results of local adaptation collective action that have benefits of low rivalry between users while being highly inclusive can be considered "local public goods". Evidence is beginning to emerge that when local governance systems facilitate high levels of participation in planning collective action for climate adaptation, and direct access to resources for implementing local plans, "local public goods" can be created and common pool resources better managed.

Les politiques générales élaborées au niveau national ne répondent pas toujours aux besoins locaux. Souvent, elles ne fournissent pas aux pauvres ruraux l'accès aux biens et aux services dont ils ont besoin pour pouvoir innover et s'adapter aux manières dont la variabilité et le changement climatiques accrus exacerbent les difficultés rencontrées pour satisfaire les besoins de base – aliments, eau, énergie et bien-être. Dans les contextes de « déficit de développement », les ressources communes sont importantes aux fins de l'adaptation au changement climatique. Pour que les actions d'adaptation au changement climatique donnent lieu à la résilience, les points de vues et les connaissances locaux doivent être reconnus et se voir accorder la priorité qui leur revient dans les systèmes de planification formels. En basant la planification formelle adaptive du développement sur les stratégies locales, on peut soutenir et renforcer des mesures que les populations ont testées et dont elles savent qu'elles fonctionnent. L'adaptation locale au changement climatique à travers des actions collectives peut permettre de lutter contre l'augmentation en cours de la variabilité climatique et les changements progressifs futurs, et satisfaire le besoin de transformer les systèmes existants pour faire face à l'évolution qualitative du climat. Ces types d'adaptation peuvent fonctionner de façons cumulatives. Les résultats de l'action d'adaptation collective locale qui présentent des avantages accompagnés de faible rivalité entre utilisateurs tout en étant très inclusifs peuvent être considérés comme des « biens publics locaux ». Des données factuelles commencent à se dégager qui indiquent que, lorsque les systèmes de gouvernance locaux facilitent un fort degré de participation à la planification d'actions collectives pour l'adaptation au changement climatique, et un accès direct aux ressources permettant de mettre en œuvre les plans locaux, des « biens publics locaux » peuvent être créés et les ressources mises en commun mieux gérées.

Las políticas desarrolladas a nivel nacional pueden resultar inapropiadas para las necesidades locales. Con frecuencia, no brindan el acceso a los recursos y a los servicios que permitirían innovar y adaptar ciertas prácticas a los pobres del campo, debido a que la variabilidad y el cambio climático exacerban los retos implicados en la consecución de los satisfactores básicos, es decir, alimentos, agua, energía y bienestar. En ámbitos subdesarrollados, resulta importante tener acceso a aquellos recursos considerados comunitarios, a fin de que sea posible implementar adaptaciones al clima. Para que las acciones orientadas a la adaptación al clima tengan resiliencia, los sistemas de planeación formales deberán reconocer e incorporar las opiniones y los conocimientos locales. En este sentido, si la planeación formal para el desarrollo adaptativo se basara en estrategias locales, podría apoyar y fortalecer las medidas que los campesinos han comprobado ya que resultan eficaces. A través de la acción colectiva, la adaptación ante el cambio climático local puede hacer frente a los actuales aumentos en la variabilidad del clima, a los futuros cambios incrementales y a la necesidad de transformar los sistemas actuales para hacer frente a los cambios cualitativos del clima. Dichas modalidades de adaptación pueden funcionar de manera acumulativa. Por otra parte, si las acciones colectivas dirigidas hacia la adaptación local se caracterizan por la reducida rivalidad entre quienes participan en ellas y si, además, son muy incluyentes, pueden ser consideradas como "bienes públicos locales". Ya existe información en el sentido de que, cuando los sistemas de gobernanza locales promueven altos niveles de participación en la planeación de acciones colectivas para la adaptación al cambio climático así como el acceso directo a los recursos necesarios para implementar los planes locales, pueden ser creados "bienes públicos locales" y los recursos comunitarios pueden ser administrados con mejores resultados.

Introduction

Local people, especially the poor and marginalised, often have little direct influence on the policies that affect their lives. Policies that are developed at national levels are seldom responsive to local needs and do not provide the rural poor with access to the assets and services they need to enable them to innovate and adapt.

In order for climate adaptation actions to deliver resilience, local perspectives and knowledge need to be recognised and prioritised in formal planning systems. Basing formal adaptive development planning on local strategies is effective as it supports and strengthens measures that people have tested and know to work. The case study presented in this paper describes a pilot project in Isiolo County, Kenya, that tests a structure for harmonising government and local planning processes through the generation of "local public goods"[1] for climate resilience. A core theme of this work has been the recognition, on the part of local people, that poor governance of common pool resources undermines adaptive capacity.

Defining climate adaptation as separate from and additional to development is impractical – particularly where development and adaptation deficits coincide. All climate adaptation must be underpinned by development that seeks to address the underlying causes of vulnerability (Ayers and Huq 2008). Development can facilitate climate adaptation and build resilience. However, development can also be largely "resilience neutral", and there can be "maladaptive development" that increases vulnerability to climate variability and change, albeit conferring other benefits, i.e., time-bound increases in productivity.[2]

Climate adaptation actions creating local public goods

Climate change is regarded as the most negative effect of the significant advances in economic development and well-being in the last century (Stern 2006). The impact of climate change on the organisation of human societies has been the subject of speculation by classical economists, political theorists, and policy analysts. Under conditions of increased resource competition conventional economic theory predicts social non-cooperation[3] (see, for example, Hardin 1968). However, empirical findings demonstrate local people's capacity to overcome such social dilemmas and create effective governance systems (Ostrom 2010).

Supporting the improvement of resource governance is regarded as a "public good" in the sense that the effectives and benefits of the better governance is non-rival and non-excludable.[4] The case study in the following section describes an approach based around the provision of "local public good" type investments in support of climate resilience.

Factors that increase the likelihood of cooperation and collective action to create public goods in the face of social dilemmas include (adapted from Ostrom 2010):

- available and reliable information about the immediate and longer term costs and benefits of actions;
- the individuals involved see the shared resources as important for their own achievements and have a longer term time horizon for rights of access and use;
- those involved have or are able to gain a reputation for being a trustworthy reciprocator;
- individuals can communicate with at least some of the others involved;
- informal monitoring and sanctioning is feasible and considered appropriate;
- social capital and leadership exist, related to previous successes in solving joint problems; and
- rules and sanctions imposed by external authorities are viewed as legitimate and enforced equitably on all.

Transposing these findings on collective responses to CPR management onto how collective action for climate adaptation might be best facilitated at the local level shows that such facilitation needs to enable the following:

- Many of those affected can agree on the need for collective action and see themselves as jointly sharing responsibility for future outcomes.
- The strengths and weaknesses[5] of climate information are well understood by local actors. Information must be tailored to their needs and accessible.
- Participants know who else has agreed to the rules and governance of the collective action and that their conformance is being monitored.
- Accessible and clear communication occurs among at least subsets of participants on a regular basis.

The case study presented below enables an analysis of how concerted adaptation action by local people can create local public goods. The hypothesis tested is that when concerted action is prioritised by the community and supported by local and national government the opportunity arises to achieve local public good climate adaptation.

The case study

The case study (drawing from Pattison, Hesse, and Anderson 2013) takes place in a largely pastoral economy of northern Kenya where customary institutions have lost much of their authority but are still recognised as the most appropriate structure for managing community resources. The authority of customary institutions for governing resource access has been compromised by the emergence of parallel, overlapping, and often contradictory systems of governance (national laws and regulations). Other broad trends among rural communities, such as greater engagement with the market economy and increasing social differentiation, have compromised the ability of customary institutions to achieve community consensus and enforce regulations. In most cases these institutions are no longer able to manage resources effectively.

The current process of government devolution in Kenya provides an opportunity to main-stream a more bottom-up approach into government planning structures. The approach described in the case study is supported by the emergence of evidence (see for example UNDP 2003; Mapesa and Kibua 2008) that when governments develop expertise in facilitating greater levels of participation, it leads to more appropriate public service provision and development interventions that address the priorities of the rural poor and climate vulnerable.

An inclusive approach to adaptive development planning is being piloted in the drylands of Kenya,[6] where mobile livestock keeping is the mainstay of the local economies. The adoption of a new national constitution in August 2010, and the realignment of local governance systems provide an ideal opportunity to transform decentralised development planning and introduce a focus on local climate adaptation. Decentralisation can provide opportunities for greater effi-ciency in the delivery of services tailored to local needs (see for example Mehrotra 2006; for the effects of decentralisation on intermediate variables affecting service delivery see Ahmad et al. 2005; Conyers 2007), better management of natural resources, and more active involve-ment of local people in the development planning system (as in Nyanjom 2011, who notes that devolution is likely to improve coherence between perceived needs and strategies for meeting them).

Under conditions of increased climate variability, improved resource governance will be required to ensure that available resources are managed in an efficient and equitable manner. Local people's capacity to adapt relies heavily on provision of appropriate and accessible public services (health and veterinary care, education, transport and communications infrastruc-ture) in addition to robust governance of the resources upon which livelihoods depend.

The approach used in Isiolo County supports bottom-up prioritisation of investments to build adaptive capacity and support resilience.[7] "Shared learning dialogues" bring a range of stake-holders together on an equal footing to discuss and analyse specific development issues. Commu-nity resource mapping that combines local and formal knowledge of landscapes and pasture and water management in GIS-generated maps proved to be a powerful medium for exchange and dia-logue.[8] In addition to building knowledge around climate change, this has fostered a greater appreciation on the part of government staff of the value of indigenous and local knowledge and the rationale behind key pastoral management strategies. It has also built the capacity of local people to articulate these issues in a way accessible to other stakeholders, and to understand the challenges faced by government staff (e.g., the constraints of budgetary cycles and mobilising resources in a timely way).

This approach was born out of a partnership with the government. In 2009 IIED was invited to work in partnership with the then Ministry of State for Development of Northern Kenya and Other Arid Lands (MSDNKOAL) to design and pilot an approach to mainstreaming climate change into planning in Kenya's drylands using Isiolo District (now Isiolo County) as the starting point. DFID provided financial and technical support. The approach developed with the MSDNKOAL is predicated on the recognition of the previous political and socio-economic marginalisation of Kenya's arid and semi-arid lands (GoK 2012).

The work in Isiolo corresponds to the way support to climate adaptation was proscribed in the Kenya National Climate Change Action Plan (GoK 2013, 227[Action 8]).

Isiolo has a population of approximately 145,000 people and a surface area of 25,000 km^2. Due to the predominant mobile livestock production system in the region, both human and livestock populations fluctuate during the year. The climate is semi-arid with an average daily temperature range of 12 to 28°C, and annual rainfall varying between 160 and 560 mm. The majority of the population are from the Borana community. Others are the Samburu, Turkana, Somali, and Meru, and a small proportion of immigrant communities from other parts of Kenya. Young people (0–14 years) account for 44.4% of the population, while the elderly (65 and above) account for 3.6%. A large proportion of the labour force (those aged 15–64 years) is either unskilled or semi-skilled and engaged in livestock-related activities (GoK 2013).

Five of the most rural wards within the larger district – corresponding to over half of the population – were chosen for the development and piloting of a bottom-up approach to climate adaptation planning. Locally prioritised adaptation investments are supported by a county-level climate adaptation fund (CAF). The CAF was initiated with £0.5 million in the first year from the UK International Climate Fund channelled through DFID. The CAF is topped up as resources are drawn down annually for local adaptation investments. The CAF approach is set out in Figure 1.

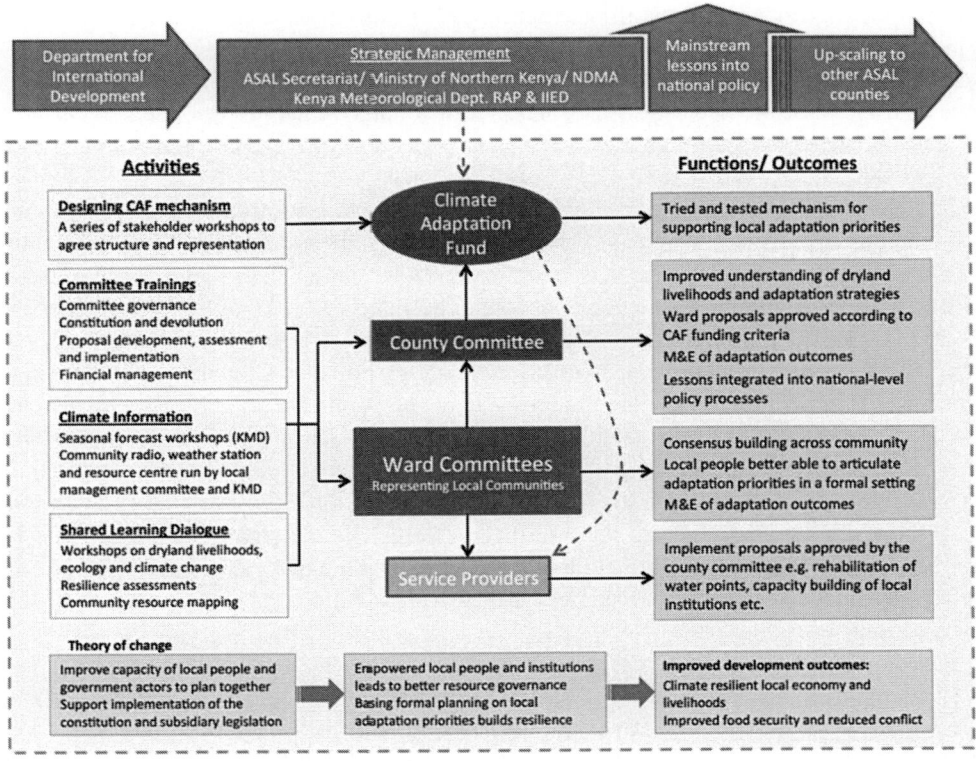

Figure 1. The climate adaptation fund approach.

The principles for the operation of the CAF were agreed with stakeholders including representatives of Wards and of local and national government. These principles include:

- The CAF supports local public good investments prioritised by communities through participatory processes.
- The CAF is managed by a county-level committee including Ward representatives, County government, and national line ministry officials.
- The CAF committee does not have the power of veto over community prioritised investments if they adhere to prioritisation criteria. The role of the CAF committee is to support, facilitate, and improve community proposals – e.g., ensuring value for money, technical feasibility, and coordination of investments within and across counties.
- Investments by the CAF must be relevant to building climate resilience.
- Investments must contribute to building harmony and peace between different communities and actors.
- Investments must support and contribute to county and national development objectives and strategies.[9]
- Investments must be viable, achievable, and sustainable.
- Investments are cost-effective and render value for money.

The CAF process can be distinguished from conventional participatory community development planning in terms of (1) the provision of climate information (seasonal forecasts and future projections) by the Kenya Meteorological Services to the local community and to local technocrats that later informed choices of interventions; (2) upstream investment in capacity building so local people were able to engage in climate-related planning; (3) the identification and prioritisation of interventions using a climate change perspective (the climate resilience assessments made in each Ward provided this focus and evidence upon which to base intervention plans); and (4) the involvement of county level technocrats to help shape viable climate-related interventions.

In 2012, ward-level climate resilience assessments were conducted.[10] These assessments used methods specifically designed to enable local people, across all ages, genders, livelihood types, and wealth statuses, to articulate the rationale underpinning their livelihood systems, and to identify solutions for strengthening their adaptive strategies and capacities. All groups highlighted the issue of improving resource governance. Inevitably, there were some areas where the priorities of marginalised groups diverged from those articulated in community meetings. Addressing these differences without causing tensions within the community was essential. A structured process of consensus-building was designed, which sought to highlight the advantages of supporting marginalised groups (e.g., women or young people) to the wider community, and ultimately led to more equitable priorities for action. The outcome of this process was a set of actions that reflected the priorities of the whole community. These actions were designed through the use of tools including scenario planning under future conditions projected from climate models. These were then used as the basis for Ward-level climate adaptation action prioritisation (see Table 1).

Based on the involvement of local and national government staff in dialogue with local people on livelihood dynamics and adaptation priorities, the pilot CAF approach to climate adaptation planning has been used to plan key actions. The first round of Ward-level planning has led to a number of projects being approved for implementation by a County-level committee of planning officials and Ward representatives. A list of the first round local adaptation projects is shown in Table 1.

Table 1.   First round projects funded by the CAF.

Ward	Project description
Merti	Construction of Bambot borehole
Merti	Capacity development for the Rangeland Users Association (RUA)
Merti	Rehabilitation of Bulesa and Muchuro farm canals
Merti	Rehabilitation of Yamicha water pan
Garbatulla	Fencing HurrBuyo water pan
Garbatulla	Strengthen natural resource management by Dedha[a] Committees
Garbatulla	Fencing Belgesh water pan
Kinna	Construction of livestock handling facilities
Kinna	Strengthen natural resource management by Dedha Committees
Kinna	Rehabilitation of veterinary laboratory
Oldonyiro	Rehabilitation and construction of sand dams
Oldonyiro	Training of water management committees
Sericho	Rehabilitation of water pans
Sericho	Rehabilitation of Hawaye wells
Sericho	Strengthen natural resource management by Dedha committees
County-level plans	Isiolo County livestock disease control programme
	Construction and establishment of locally managed radio station for the county in order to deliver climate and weather information, etc

[a]*Dedha* refers to the Borana customary institution for management of natural resources. As noted above, these institutions retain broad community support while in practice being redundant in some cases.

The selection of Ward-level CAF committees was a very transparent and well publicised process that incorporated public scrutiny of each candidate based on their record of serving the community and their skills (this process is referred to as "public vetting"). In part this process was motivated by a desire on the part of local people to move away from reliance on the usual suspects who sit on many different committees and are able to position themselves as gatekeepers to community participation (with the personal benefits and influence that such a position provides).

Despite the decline in the authority of customary resource management institutions described above, these institutions do still offer a model for managing natural resources that retains legitimacy in the eyes of the vast majority of local people. With the exception of the Rangelands Users Association (RUA) in Merti Ward (which with donor support has transformed itself from *Dedha* Committee into a relatively effective hybrid *Dedha*/NGO organisation), most *dedha* committees are ineffective at managing resources for the benefit of the community at large and require improvements in capacity in parallel with formal recognition by County government. Without the existence of these institutional structures the support required under CAF to improve resource governance would have been significantly larger and more time consuming.[11]

Analysis of the process

The case study was analysed using qualitative assessments of the public good attributes of the climate adaptation categories arising from the case study planning process and the conditions achieved in the case study to engender collective action. The results are presented in Tables 2 and 3.

Table 2.   Assessment of the public good attributes of the climate adaptation categories arising from the case study planning process.

Categories of climate adaptation actions	Public goods attributes of adaptation measures	
	Rivalry of use	Exclusion from use
Natural resource access measures (new or rehabilitated)	Low – more efficient resource use such that climate-related scarcity is reduced	Low – rights-in-use for access determined by customary institutions
Protection of natural resources	Low – intended to conserve resources in the face of degradation	Low – rights-in-use for access determined by customary institutions
Human capacity development of post holders in customary institutions	Medium – candidates for capacitation are representatives from the community. They receive private benefits as well as providing public good	Low – benefits of better performance by institutions should be open to entire community
Physical infrastructure (creation or rehabilitation)	Low – public infrastructure to increase resource accessibility and reduce rivalry	Low – utility of public infrastructure subject to asset holdings
Public service provision	Low – provision according to demand	Low – intended to be free at the point of access

Table 3.   Conditions achieved in the case study to engender collective action.

Conditions for adaptation collective action	Attributes of the case study – all activities involve representatives of marginalised groups in local communities including women-headed households, poorest, young, and elderly
Stakeholders agree on need for collective action and shared responsibility for outcomes	Climate resilience assessments carried out with communities. Co-design of local climate adaptation planning and implementation processes. Governance agreed and local committees established through a transparent nomination and vetting procedure
Regular and accessible information about the effects of increased climatic variability and change	Kenya Met. Dept. provides seasonal forecasts, climate change awareness information and early warning information of extreme events. A local radio station is being established to disseminate this information more widely and in a timely way
Stakeholders know who has agreed to governance of collective action and conformance monitored	"Town hall" meetings (barazas) held where representatives of communities and members of Ward planning committees can discuss progress and priorities
	Local NGO monitors process in all Wards and issues reports to stakeholders. Information will be broadcast on local radio
Accessible and clear communication among stakeholders on a regular basis	Regular meetings held at Ward and County levels to socialise information on progress of processes

Initial lessons

The action research carried out in Isiolo County during the establishment of a local climate adaptation planning and implementation process has been documented (Hesse and Pattison 2013) and subjected to a collective analysis and reflection process among the partners involved. The main lessons drawn from the experience so far are described below.

Awareness creation

Local people were provided with information on the increasing climatic variability and longer-term change, and the implications for planning. Seasonal forecasts were presented and discussed at Ward meetings in terms of what they meant for different livelihoods.

Moving toward climate adaptation from development deficit

Planning climate adaptation in a development deficit situation reveals the compounded nature of development needs with adaptation priorities. As local climate adaptation planning becomes institutionalised into the formal development planning cycle, so the distinctions and complementarities among development and climate adaptation can become clearer to local people and formal planners.

Local/formal links

The roles of local, customary, and informal institutions in bottom-up planning for adaptation and resilience are vital. Local knowledge on adaptive practices, particularly on the management of natural resources, should be central to planning.

Local plans can be aggregated across households, groups, enterprises, and communities, and then proposed as components of local development plans at the county and national scale. In this way local priorities and strategies are mainstreamed first into decentralised and then national development planning.

For inclusive policy-making based on the priorities and knowledge of local people there must first be some level of shared understanding of the key issues with local government staff. Elected officials have a role in representing community concerns and priorities to the county assembly but in terms of planning local public good investments, local government must have a mechanism for engaging directly with local people. Sustained dialogue over extended periods of time allows a range of stakeholders adequate time for reflection and learning. Change – transformative change – does not happen overnight.

Institutionalisation as an imperative and funding access

Building a more coherent development planning system cannot be done through the standard "projectised" approaches used by NGOs that create parallel processes and structures to the detriment of the development of local government capacity. Transformative climate adaptation requires that local government is pivotal to the process. The sustainability of an effective planning and policy process ultimately depends upon this.

The process in Isiolo started before county governments were in place but nevertheless worked with existing structures – district government officials and the county council. The adaptation planning committees drew membership from state and non-state actors and worked closely with District Development Committees, District Steering Groups, etc. A lot of effort is now being made to institutionalise the CAF and related planning processes into the County governance architecture.

For effective local to national policy development, funds must be channelled through demand-side determined measures in national development systems that enable local people to identify their development needs and address their climate vulnerabilities. It is for this reason that mainstreaming the CAF approach into County government planning processes is the goal of the pilot. The intention is that CAF will not exist as a parallel structure beyond the pilot phase. Strong government engagement at the County and National-levels will ensure that lessons learnt are both mainstreamed and up-scaled to other arid and semi-arid counties.

Local public goods climate adaptation

The concept of climate adaptation through support for local public goods is accessible to stakeholders and useful in framing dialogue toward local climate adaptation planning and implementation where local people are involved in and/or support customary governance systems for shared common pool resources management.

Scale of planning for climate adaptation

The CAF approach emphasises that planning for climate adaptation needs to occur at appropriate scales rather than being restricted to administrative boundaries (for example, planning for domestic water supplies may be done at the village level whereas planning for livestock mobility and water point governance may need a county or cross-county approach). This is sometimes referred to as landscape-level planning and necessitates a structure for county planners to coordinate across county and even national borders. As CAF expands into neighbouring counties, this will be an important element to develop and test further (Hesse and Pattison 2013).

Challenges

Finding local partners with whom all groups within the wider community, customary institutions, and local government are happy working can be challenging. A significant upstream investment in building inclusive dialogue around this issue drew out potential tensions and allowed the formation of strategies to minimise friction between different groups. There is a real danger of partial community participation if local partners are not fully representative.

Conclusions

Analysis of the experience of establishing planning and implementation processes for local climate adaptation, based on county-level CAFs in northern Kenya leads to the following conclusions:

- Local collective action can be concerted on climate adaptation through a combination of stakeholder consultation, participative climate resilience assessments, regular climate information provision, resource mapping across landscapes to identify climate resilience attributes, and establishing mechanisms for drawing down resources to cover the investments costs of adaptation actions.
- Local public goods can be created through local climate adaptation planning and implementation in situations where community-supported customary institutions can be included in the process.
- Initial local adaptation priorities[12] will inevitably address the development deficits that impair climate adaptive capacity to perceived increases in climatic variability.
- If such collective actions are recognised by local government as legitimate contributions to the formal planning process (on an iterative basis) local adaptation can be transformative.

The approach used in the case studied here for supporting local climate adaptation looks both at addressing risks due to increased climatic variability, and by doing so, the opportunities to improve formal development planning at decentralised and national scales. The success of the work in Isiolo and elsewhere has led to it being up-scaled through an "Adaptation Consortium", where a hybrid of similar approaches is being generated and implemented across four other semi-arid counties of Kenya. This unified approach works across institutional barriers and allows comparability in delivery costs by different implementing partners.

Notes

1. "Local public goods" refers to processes and resources that are largely non-rival (use of the resource or investment does not diminish that amount left for other potential users) and non-excludable (use of the resource or investment is not restricted to certain people).

2. In practice these conceptual categories are rarely discrete. However, the categories do provide a useful summary of how adaptation can differ from, or be additional to, "business as usual" development.

3. Where uncoordinated decisions motivated by the pursuit of individual benefits generate suboptimal payoffs for others and for self in the long run.

4. Goods that are both non-rival (a rival or "subtractable" good is one where consumption by one user does not prevent simultaneous consumption by another) and non-excludable (no one is prevented from consuming) are called public goods. It is generally accepted by mainstream economists that the market mechanism will under-provide public goods, so these goods have to be produced by other means. "Local public goods" refer to rivalry and exclusion among a specific community or stakeholder group. Both "rivalry" and "exclusion" are considered continuous variables.

5. The reliability of seasonal forecasts is reduced by poor weather station coverage (Africa has only one weather station per 26,000 km^2 – one-eighth the recommended minimum). Climate projections concerning likely conditions one or several decades into the future are considered of moderate reliability. Flooding forecasts based on rainfall further up the catchment can be considered highly reliable and can trigger emergency responses that protect key assets.

6. Building on the LAPA framework developed in a partnership among the Government of Nepal, DFID, and IIED (see Government of Nepal 2011).

7. This approach differs from the majority of participatory planning which in practice is often no more than a process of consultation by government or NGOs to legitimise their own programmatic focus. Through a process of shared learning dialogue, local people not only prioritise their own adaptation investments but they also monitor and evaluate implementation.

8. Ced Hesse "Responding to climate change in East Africa by strengthening dryland governance and planning." Accessed November 10, 2013. http://www.iied.org/responding-climate-change-east-africa-strengthening-dryland-governance-planning.

9. This initiative is piloting an approach to be mainstreamed into wider government planning processes. Hence, it is essential that investments align with government policy at all levels.

10. Climate resilience assessments for the Wards of Kinna, Sericho, Gafarsa, Merti and Oldonyro were carried with representatives of the residents in 2012. Summaries of the resilience assessments can be found at http://pubs.iied.org/pdfs/G03464.pdf; http://pubs.iied.org/G03467.html; http://pubs.iied.org/G03468.html; http://pubs.iied.org/G03465.html; and at http://pubs.iied.org/G03466.html. Accessed October 1, 2013.

11. The Isiolo CAF was initiated before County governments came into being in mid-2013. But the CAF establishment process worked with the District Council and local government officials. The Council allocated 10 acres of land for building the community radio and weather observation station in Garbatulla; the County government allocated Ksh10million (€82,990) to building staff houses. Ward committees made submissions during preparation of County Integrated Development Plan based on the Ward resilience assessments. Perhaps the more important than the tangible investments has been the sense of ownership built within Isiolo among both officials and community members:

"Our Isiolo map used to be plain with very little features of dots representing small towns. This time thanks to RAP and their partners our Isiolo map is full of resources ... We will incorporate this great work as the Isiolo county resource map and give all the support it requires." (HE Hon. Godana Doyo, Governor, 7 November 2013)

12. Addressing the development deficits revealed by the effects of increased climatic variability is considered to be "adaptive development". This differs from "business as usual development" in that the likely impacts of climate variability and change are given due consideration in planning investments such that maladaptive actions are avoided and steps are taken to climate proof development.

References

Ahmad, J., S. Devarajan, S. Khemani, and S. Shah. 2005. *Decentralisation and Service Delivery*. Policy Research Working Paper 3603. Washington, DC: The World Bank.

Ayers and Huq. 2008. Adaptation Funds and Development Assistance: Some Frequently Asked Questions. IIED Sustainable Development Briefing paper. November 2008.

Conyers, D. 2007. "Decentralisation and Service Delivery: Lessons from Sub-Saharan Africa." *IDS Bulletin* 38 (1): 18–32.

GoK (Government of Kenya). 2012. *Sessional Paper No.8 of 2012 on The National Policy for the Sustainable Development of Northern Kenya and Other Arid Lands*. Nairobi: Government of the Republic of Kenya.

GoK. 2013. *National Climate Change Action Plan 2013-2017*. Nairobi: Government of the Republic of Kenya.

Government of Nepal. 2011. *Adaptation to Climate Change: NAPA to LAPA*. Kathmandu: Ministry of Environment, Government of Nepal. Accessed 10 January, 2014. http://moste.gov.np/policy_documents/strategies#.UZcBX0okRht

Hardin, G. 1968. "The Tragedy of the Commons." *Science* 162 (3859): 1243–8.

Hesse, C., and J. Pattison. 2013. "Ensuring Devolution Supports Adaptation and Climate Resilient Growth in Kenya" (*IIED Briefing*, June, 2013). Accessed 10 January, 2014. http://pubs.iied.org/17161IIED

Mapesa, B., and T. N. Kibua. 2008. "Management and Utilization of Constituencies Development Fund in Kenya". In *Decentralisation and Devolution in Kenya: New Approaches*, edited by T. N. Kibua and G. Mwabu, 236–251. Nairobi: University of Nairobi Press. Chapter 12.

Mehrotra, S. 2006. "Governance and Basic Social Services: Ensuring Accountability in Service Delivery Through Deep Democratic Decentralisation." *Journal of International Development* 18: 263–83.

Nyanjom, O. 2011. *Devolution in Kenya's New Constitution*. Constitution Working Paper No. 4. Society for International Development (SID). Accessed 10 January, 2014. http://www.sidint.net/docs/WP4.pdf

Ostrom, E. 2010. "Beyond Markets and States: Polycentric Governance of Complex Economic Systems." *American Economic Review* 100 (3): 641–72.

Pattison, J., C. Hesse, and S. Anderson. 2013. "Local to National: Putting Local Voices at the Heart of National Policies." Paper presented at the "Hunger, Nutrition and Climate Justice: A New Dialogue: Putting People at the Heart of Global Development" conference, Dublin, 15–16 April.

Stern, N. 2006. *The Stern Review on the Economics of Climate Change*. Cambridge: Cambridge University Press (HM Treasury). http://webarchive.nationalarchives.gov.uk/+/http:/www.hm-treasury.gov.uk/sternreview_index.htm

UNDP. 2003. *Fiscal Decentralisation and Poverty Reduction*. Washington, DC: UNDP.

Sustainable rural livelihoods approach for climate change adaptation in Western Odisha, Eastern India

Virinder Sharma, Bhaskar Reddy, and Niranjan Sahu

The economy of Odisha is primarily agrarian. Over 80% of the population of Odisha live in rural areas, where levels of poverty are higher than in the state's towns and cities. They depend for their livelihoods on farming and collecting forest products. During the dry season, many migrate elsewhere in Odisha and nearby states in search of temporary work as labourers. Odisha has the highest proportion of inhabitants from scheduled tribes and scheduled castes of all the states in India (39.9% compared to 24% nationally). These groups are marginalised and experience high rates of poverty, low levels of education and poor health. They are highly vulnerable to climate change, due to poverty and dependence on climate-sensitive livelihoods in a vulnerable region. The Western Odisha Rural Livelihoods Project sought to reduce poverty by improving communities' water resources, agriculture, and incomes. Communities were involved throughout and are now better able to respond to climate variability (both droughts and heavy rains). The Government of Odisha took full ownership of the project and state and national governments subsequently adopted approaches used by WORLP.

L'économie de l'Odisha est principalement agraire. Plus de 80 % de la population de l'Odisha vit dans des zones rurales, où le niveau de pauvreté est plus élevé que dans les petites et grandes villes de l'État. Ces personnes sont tributaires, pour ce qui est de leurs moyens de subsistance, de l'agriculture et de la collecte de produits forestiers. Durant la saison sèche, nombre d'entre elles migrent vers d'autres parties de l'Odisha et des États voisins en quête de travail temporaire comme ouvriers agricoles. L'Odisha affiche le plus grand pourcentage d'habitants issus de tribus et de castes répertoriées de tous les États indiens (39,9 % par rapport à 24 % à l'échelle nationale). Ces groupes sont marginalisés et affichent un fort taux de pauvreté, un faible niveau d'éducation et une santé médiocre. Ils sont extrêmement vulnérables au changement climatique, parce qu'ils sont pauvres et tributaires de moyens de subsistance sensibles au climat dans une région elle-même vulnérable. Le Western Odisha Rural Livelihoods Project (WORLP) a cherché à réduire la pauvreté en améliorant les ressources en eau, l'agriculture et les revenus des communautés. Les communautés ont pris part à tous les efforts et sont maintenant plus à même de faire face à la variabilité climatique (tant les sécheresses que les fortes pluies). Le gouvernement de l'Odisha s'est pleinement approprié le projet et les autorités gouvernementales nationales et d'autres États ont ultérieurement adopté des approches utilisées par le projet WORLP.

La economía de Odisha es principalmente agraria. Más de 80% de su población vive en zonas rurales, en las que los niveles de pobreza son más elevados que en las áreas urbanas del estado. Sus medios de vida dependen de la agricultura y de la recolección de productos forestales. Durante la temporada de sequía, gran parte de la población migra hacia otras localidades de Odisha y hacia estados cercanos, buscando trabajo temporal como obrero. Considerando todos los estados de India, Odisha registra la proporción más elevada de habitantes de tribus y de castas reconocidas (39.9 % *vs*. 24 % a nivel nacional). Dichos grupos se encuentran marginados y muestran elevadas tasas de pobreza, bajos niveles de escolaridad y una salud precaria. Asimismo, como consecuencia de su pobreza y su dependencia acusan muy alta vulnerabilidad ante el cambio climático en medios de vida sensibles al mismo en una región vulnerable. El Proyecto de los Medios de Vida Rurales de Odisha Occidental se orientó a reducir la pobreza a través del mejoramiento de los recursos de agua, de las prácticas agrícolas y del aumento de los ingresos a nivel comunitario. Las comunidades participaron en todas las fases del proyecto y hoy se encuentran en mejores condiciones para da respuesta a la variabilidad climática, es decir, a las sequías y a las tormentas fuertes. El gobierno de Odisha participó plenamente en el proyecto y, posteriormente, otros gobiernos, tanto a nivel estatal como nacional, incorporaron los enfoques utilizados en el mismo.

Introduction

The economy of Odisha, one of the Eastern states in India is primarily agrarian. Over 80% of the population of Odisha lives in rural areas, where levels of poverty are higher than in the state's towns and cities. Odisha has the highest proportion of inhabitants from scheduled tribes (ST) and scheduled castes (SC) of all the states in India (39.9% compared to 24% nationally). These groups are marginalised and experience high rates of poverty, low levels of education, and poor health. Four of the poorest Western Odisha districts, i.e., Bargarh, Bolangir, Nuapada, and Kalahandi were selected for implementation of Western Odisha Rural Livelihoods Project (WORLP) at the initial stage. Bolangir and Bargarh districts are a part of the "West Central Table Land Zone" and have a hot and sub-humid climate.[1] Similarly, Kalahandi and Nuapada districts are a part of the "Western Undulating Zone" and have hot moist and sub-humid climate.[2] Both agro-climatic zones are located in Eastern Plateau and Hills Zone (Zone number 7) of India (Behera et al. 2005). People living in this region are likely to be witnessing deteriorating climatic conditions (floods, drought, and temperature rise), with increased risks from disease and pests, and with associated implications for human and livestock health. Over half of the people living in western Odisha are from ST and SC communities. These people in western Odisha are highly vulnerable to climate change, partly as their poverty limits their capacity to deal with shocks and stress, partly as a result of living in an area of high environmental risk with dependence on climate sensitive livelihoods. This is a region of India where the mean temperatures are rising, and where the vulnerability profile places it among the highest risk areas in the country (WORLP 1999).

Western Odisha rural livelihoods project (WORLP)

Odisha implements around 10 different watershed programmes and projects in the state through the Odisha Watershed Development Mission (OWDM), an autonomous state agency constituted under the Department for Agriculture that plans, implements, and monitors watershed development programmes in the state. WORLP is one of these programmes of the Odisha government and is funded by the United Kingdom's DFID and implemented over a period of 10 years, i.e., from 2000 to 2011. The cost of the project is Rs. 2300 million (£32.75 million representing 18% of DFID's total spending of £183 million in Odisha). The project initiated a new approach to watershed management, termed "Watershed Plus". It provides a range of livelihood support services to the poor. WORLP was designed using the sustainable livelihoods approach, which provides a conceptual and methodological framework for addressing poverty (DFID 1999).

The goal, purpose, and outputs of the project, as described in the project framework, are as follows:

- Ultimate goal: Reduce poverty in rain-fed areas of India.
- Intermediate goal: By the end of the project year (2010) government agencies and other stakeholders in Koraput, Bolangir, and Kalahandi districts and elsewhere adopted more effective approaches to sustainable rural livelihoods.
- Purpose: Promote sustainable livelihoods, particularly for the poorest, in four districts (Bolangir-14 blocks, Nuapara-5 blocks, Bargarh-4 blocks, and Kalahandi-6 blocks) in replicable ways by 2010.

The project's planned outcome was to "*promote sustainable livelihoods, especially for the poorest people*", in four districts, "*in replicable ways*", by 2011 (WORLP 1999). It was able to cover 1180 villages in 677 watersheds in these four districts, where human development indicators are very low. The project originally aimed to support 125,000 households, comprising 625,000 people living in 290 watersheds. The WORLP project area is hilly and the project focussed on watersheds of approximately 500 hectares. In 2006, it was decided to extend the "Watershed Plus" activities to another 387 "additional" watersheds, where the government had already improved land and water management under its regular programme. Figure 1 outlines the basic shape of land and water management in WORLP.

WORLP Interventions: The major interventions can be broadly categorised as:

- Land and water management: Land and water management called the "Watershed development" (see Box). This comprised activities to improve the management of land and water, including putting in place infrastructure such as embankments, water storage ponds, and irrigation channels, etc.
- Economic support for the poorest: Economic support for the poorest called the "Watershed Plus" component. Activities under this component included providing loans and grants for micro enterprises and microcredit and ensuring access to common land for joint enterprises.

A watershed is an area of land drained by a river or stream. Watersheds can vary in size from small valleys drained by a single stream to large river basins like the Ganges.

The WORLP project area is hilly and the project focussed on watersheds of approximately 500 hectares. Within these, WORLP supported the integrated development of land and water resources from the hills and ridges bounding the watershed down to the valley floor.

On the upper slopes, communities typically planted orchards and other trees to reduce run-off and provide incomes for common interest groups of poor people. At the bottom of the slopes, communities dug ponds and embankments to slow the water rushing off the hills in heavy rain. On the lower lands, Watershed groups dug water storage ponds and irrigation channels to irrigate the rice crop at the end of the monsoon season and other crops in the dry season. Sometimes, small concrete structures like sluice gates were also built.

Figure 1.   Land and water management in WORLP.

- Capacity building: Capacity building component was used to empower communities and facilitators resulting in adoption of better practices for natural resources management and livelihoods promotion.

Climate change adaptation and building resilience

Climate change and variability are among the most important challenges facing developed countries because of their strong economic reliance on natural resources and rain-fed agriculture. People living in marginal areas such as dry lands or mountains face additional challenges with limited management options to reduce impacts. Climate adaptation strategies should reflect such circumstances in terms of the speed of the response and the choice of options. In view of the above, a framework for climate change adaptation needs to be directed simultaneously along several interrelated lines (FAO 2007):

- **Legal and institutional elements** – decision-making, institutional mechanisms, legislation, implementing human rights norms, tenure and ownership, regulatory tools, legal principles, governance and coordination arrangements, resource allocation, networking civil society.
- **Policy and planning elements** – risk assessment and monitoring, analysis, strategy formulation, sectoral measures.
- **Livelihood elements** – food security, hunger, poverty, and access.
- **Cropping, livestock, forestry, fisheries, and integrated farming system elements** – food crops, cash crops, growing season, crop suitability, livestock fodder and grazing management, non-timber forest products, agroforestry, aquaculture, integrated crop-livestock, silvo-pastoral, water management, land use planning, soil fertility, soil organisms.
- **Ecosystem elements** – biodiversity/species composition, ecosystem services.
- Linking climate change adaptation **processes/technologies** for carbon sequestration, fossil fuels substitution, and promotion of bioenergy.

Watershed plus approach

The typical nature of implementation as a land-based programme manages key resources such as land, water, and forest resources. The interventions are decided primarily on the occurrence, distribution, and management of precipitation. The benefits flow to the well-off and those who owns land. The watershed guidelines do not include provisions for benefiting the landless and asset-less. WORLP operated in a watershed platform in rain-fed areas with natural resources management as its core component. It also addresses the issues of institutions, decentralised governance, capacity building, livelihoods promotions, and ecosystem services. The results of WORLP were analysed to assess its impacts on community resilience to climate change. Both the government of Odisha and India were keen to reform the programme to achieve greater poverty impact. Contributing to this effort, during the design phase, it was decided to provide an appropriate framework for rural livelihoods that has a greater overall impact and sustainability.

This was done by adding a livelihood component to the classical watershed programme. This additional component, with additional activities and budgetary provisions was termed as "watershed plus". WORLP was the first project under this new framework. The project made a conscious effort to build five types of capital, i.e., Natural, Physical, Financial, Social, and Human capital. This led to the shift in the focus of mainstream watershed programme towards the landless and asset-less. These increases in five types of capital enable the community adapt to the effect of climate change. Thus watershed plus approach included three major components to address the weaknesses of the old approach, which are: promoting livelihood improvements; capacity building for primary and secondary stakeholders, and encouraging an enabling environment.

Design and methods

This paper attempts to test the hypothesis that the watershed plus approach builds resilience of communities to climate change through promotion of sustainable rural livelihoods. Watershed programmes by nature focus on natural resources management. The results are based on a comprehensive Climate Change Adaptation study in 2009 (Satyanarayana et al. 2009) and the Independent evaluation (post project) of WORLP in 2011 (WORLP 2011). The Impact assessment study was conducted by independent agencies (Sambodhi and Winrock International).

Climate change adaptation study (2009)

A study documented the ways WORLP area communities coped with and adapted to increased climate-induced vulnerability. The study identified ways people in the project area are adapting to climate change and proposed ways to make livelihoods more resilient. In summary, the study was intended (1) to understand adaptation, coping, and responses of the project community to climate variability; (2) to identify climate resilient technical measures that might be replicated and help in promoting sustainable livelihoods; and (3) to identify and recommend socially acceptable adaptation and mitigation options for the project (Satyanarayana et al. 2009). The indicators used in the analysis of this study are given below. Wherever possible, comparisons are made between ex-post and ex-ante project scenarios:

- Socio-economic factors and anthropogenic effects on land use change;
- Natural resource management interventions and their effect on soil moisture;
- Effect on crop production and crop intensification cycles;
- Social capital and community resilience for adaptation;
- Strategies for coping and community responses to weather events;
- Carbon cycle and mitigation efforts;
- Macro and micro models using national climate data and local data (soil, rainfall, temperature) for assessing variability in bio-physical factors.

WORLP lies in an area of India where the mean temperatures are rising, and where the vulnerability profile places it among the highest risk areas in the country (Satyanarayana et al. 2009). The spatial patterns of linear trends in temperature in India from 1901 to 2000 shows that Odisha and WORLP areas lie within an area that is warming (Satyanarayana et al. 2009, 9, section 2.4). The following climate risks have been identified in the project area:

- High variability of rainfall, leaving people with two peak periods of food stress;
- Droughts and dry spells every two years, with a major drought every five to six years;
- Flash floods during the rainy season.

There are different ways to address climate change phenomena: (i) through mitigation measures, (ii) through adaptation processes, or (iii) through coping. Mitigation measures are long term and require constant effort and global consciousness. Adaptation through various coping mechanisms to build in resilience towards climatic variability is an ongoing process, focusing on sustainable management of natural resources and livelihood diversification. Although mitigation and coping are addressed here, this paper primarily discusses the ways people in the project area are adapting to climate changes, and the project's contribution in making livelihoods more resilient.

Potential linkages of WORLP to climate change

Climate change adaptation in WORLP comprises of a range of measures and initiatives that reduce the vulnerability of human and natural systems to climate change. It was not designed with any climate change objectives, and no environmental impact was envisaged, other than the enhancement of natural resource assets. Nonetheless, the project has increased the asset levels of the poor and very poor, which in turn has ensured that they are better able to cope with anticipated hazards and adapt to a changing environment by building their resilience. The five project outputs are tabulated below, with comments on their potential link with climate change Figure 2.

Vulnerability, poverty reduction, and gender

The sustainable livelihoods approach, illustrated and guided by the framework, was very much at the core of the design process. It hypothesised that the enhancement of people's livelihood assets would, given supporting processes, lead to the development of effective strategies and eventually to positive livelihood outcomes. But it always recognised that these processes were prey to what was termed the vulnerability context, where shocks, adverse trends, and seasonality – over which they had no control – had the capacity to drag people back into poverty. In many ways, this model has proven to be a powerful one in the light of more recent climate change evidence, and the sustainable livelihoods approach hypothesis has been to a large extent confirmed.

WORLP districts have a high level of vulnerability because they have a lot of poor people (the population density of Odisha is 269/km^2, with 17.1% scheduled caste and 22.8% scheduled tribe people) living in a reduced natural environment (Census of India 2011). There is very weak asset base. Access to natural, financial, and human capital is restricted. High levels of unemployment

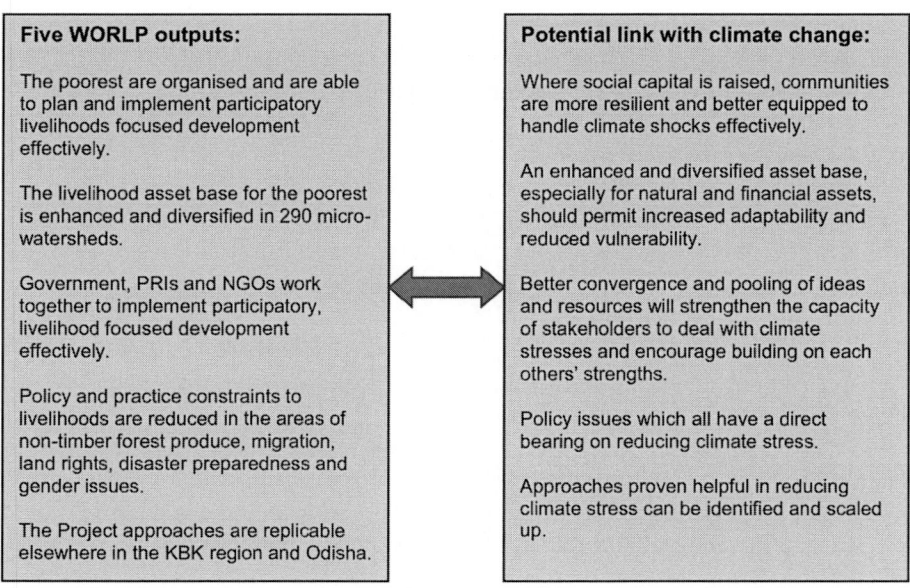

Figure 2.   Potential link of project outputs to climate change.

and low levels of literacy exacerbate the situation. In the partnership between the communities, the project, and government staff, engaged in participatory micro-planning processes with the full engagement of poor people. This helped to build trust and good relations, identifying the needs and concerns of those most at risk. The approach has created an enabling environment, empowering and informing people, and allowing them to make informed choices for their long-term well-being. As a result, in the areas where WORLP was implemented, the incidence of poverty reduced during the project duration by 30%, meaning that approximately 15,000 households or 72,000 people have moved above the Govt. of India-defined poverty line. Much of this can be attributed to increased levels of financial, human, natural, and social assets brought in under the project, assets which also have built resilience, and improved adaptability to climate change.

Increased community resilience has developed through the project's efforts to build the capacity of individuals, households, and groups that face multiple environmental and other pressures, and in ensuring their increasing control over resources. Farmers have been supported to increase their skills related to crop diversification, vegetable gardens, aquaculture, duckery, and goatery, etc. This increase in skills enables people to adapt their livelihoods and build resilience to climate changes and shocks. This has also been supported by increases in health and well-being through health camps, water and sanitation initiatives, and the introduction of smokeless *chullahs* (firewood stoves) and other technologies, such as surface treadle pumps. The increases in skills and opportunities have been evidenced by a decrease in stress-induced migration from 47% of households in 2000 to less than 15% in 2008 (WORLP 2008). It is the poorest of the poor who constitute the distress migrants. Previously they were unable to find work in the *rabi* season when water is scarce.

Project initiatives also had positive effects on women, as it increased women's capacity to adapt to climate induced stress and to cope in situations of crisis. This has to a large extent happened through the strengthened resilience brought about through a large number of capable self-help groups, where women are able to share resources and ideas, and thus reduce their inherently high levels of vulnerability to disasters caused by climate change. Migration and the associated stress, which is particularly acute in the case of women, have been very substantially checked by project activities, from almost 50% incidence of migration to under 15% (WORLP 2011). In addition to these positive effects of increased social capital, some of the effects of enhanced natural capital may be seen as favouring women, such as improved food security, improved health status including child nutrition, and reduced drudgery.

Institutional arrangements and WORLP project structure and social capital

In order to analyse the state's response to extreme climate events, it helps to understand the institutional arrangements within the Odisha Watershed Development Mission (OWDM) and WORLP. These institutional arrangements shape and inform the potential mechanisms for coping with climate change. The institutional set-up within WORLP, driven by the Odisha Watershed Development Mission, has allowed a high level of autonomy and flexibility. Activities that appear to increase people's capacity to adapt to and cope with climate-related stress have been implemented in a quick, effective, and participatory way, and through a direct chain of command. This has allowed the project institutions, which were designed to operate in a highly participatory mode at all levels, to operate more effectively.

At the state level, WORLP is under the aegis of the Director of the OWDM. At the district level, the Project Director (PD) watersheds located in the project districts have independent offices. Project implementing agencies and watershed development teams implement the project in a specified number of watersheds at the block level. The implementation and governance functions are segregated at each level by the empowered committee at the state level and the district watershed development committee at the district level. WORLP has a structure at the district level that supports the implementation of the watershed and watershed plus activities. The district project director manages the watershed activities in a district with the support of assistant project directors. The project director's office also has the capacity building team consisting of four members specialising in livelihoods, micro enterprise, natural resource management, and monitoring and evaluation.

At the block level, the activities are coordinated by project implementing agencies. The Project implementing agencies could either be an NGO or a government staff. A three-member livelihood support team made up of specialists from the agriculture, micro enterprise, and social development sectors are attached to the project implementing agency to support the implementation and monitoring of watershed plus activities in the intervention areas. The social development livelihood support team is responsible for gender and community organisation activities. The frontline contact is made by the watershed development team at the village level with support of community link workers, village volunteers who work closely with the community. The community link workers act as bridge between the project and the community and provide services on agriculture, natural resources management, livestock, and community mobilisation activities. At the watershed level, WORLP has an environment that encourages the participation of women, the poor, and other marginalised groups in watershed activities. The institutional arrangement in WORLP is given in Figure 3.

At each level (state, district, block, and village), WORLP has created parallel support structures for capacity building, enabling policy formulation, and facilitating project processes. Thus, the creation of project support units, capacity building teams, and livelihood support teams was an innovation that has helped the project to achieve its purpose. The links and roles of institutions in the project that have a bearing on climate change related action are described in Table 1. Some of these roles have been performed in the past and some are desired for the future.

Strategic decisions on adaptation and convergence are taken by the Director, OWDM. Most responses to the climate crisis are coordinated by the collector in the district. Project implementing agencies take operating decisions by taking watershed associations and self-help groups into confidence for adaptation and coping. Some project implementing agencies also work with district

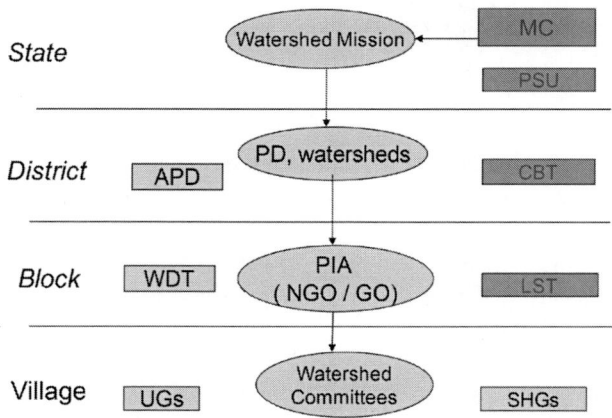

MC=Managing Consultants, PSU=Project Support Unit, CBT=Capacity Building Team, PIA=Project Implementing Agency LST=Livelihood Support Team, SHGs=Self Help Groups, APD=Assistant Project Director, WDT= Watershed Development Team, UG=User Group

Figure 3.   Institutional arrangement in WORLP.
Source: Satyanarayana et al. 2009.

Table 1.   The links and roles of institutions in WORLP bearing on climate change.

Input/activity	Institution responsible	Possible linkages	Remarks
Overall linkage and awareness about climate events	OWDM	Adaptation Mitigation Coping	Even though not done consciously, the direction of OWDM has a direct bearing on reduction of climate stress
District-level planning and convergence	Project directors	Adaptation Mitigation Coping	Project directors plan activities in consultation with line departments and provide platform for convergence
Capacity building on aspects of water harvesting structures	CBT-NRM, Project implementing agencies, WDTs	Adaptation and Coping	Soil moisture conservation help in adaptation as well as coping
Capacity building on aspects of cropping systems	CBT-NRM LSTs and WDTs	Adaptation	Advice on cropping system and moisture management through intercrop
Capacity building on aspects of land use, irrigation, plantation	LSTs, Project implementing agencies, WDTs	Adaptation	Land use change and varietal change due to change in moisture
Animal health care	CLWs, Project implementing agencies	Response	Better preparedness
Social capital building	Self-help groups, CIGs	Response Coping	It has a strong linkage in risk mitigation and coping

authorities on responding to climate induced events and other critical events. Capacity building teams provide strategic coordination for adaptive actions, and livelihood support teams have the same role, but with a more hands-on approach in watershed areas. The project has made substantial investments in capacity building of project beneficiaries. The social capital of the community was thereby increased. This is possibly the project's greatest contribution to increasing people's capacity to adapt in crisis. Greater social capital is likely to increase people's resilience and capacity to cope. They will be better informed and able to make more appropriate responses to stress situations. The increased number and strength of self-help groups has increased the stock of social capital within the project, and this has had an immediately reduced people's vulnerability to the negative effects of climate change. Through the groups' exposure to participatory planning processes, people are better able to manage common property resources and are more prepared for crises than those in areas where such groups are non-existent or weak.

In summary, the institutional set-up within WORLP, driven by the OWDM, has allowed a high level of autonomy and flexibility. The structure at every level has maximised opportunities for increasing people's participation. Activities which appear to be very beneficial in terms of increasing people's capacity to adapt to and cope with climate-related stress have been implemented in a quick, effective, and participatory way, and through a direct chain of command. This had allowed the project institutions to operate effectively and substantially improve project delivery.

Environment and food security

There is some evidence of changing trends in the project area's micro climate over recent decades. There has been up to 10% more rain in the monsoon periods (although it has been more variable), and less outside of it. As a result, the likelihood of prolonged dry spells or drought has declined. The OWDM has undertaken many interventions since 2001, and many watershed projects have completed their project period of five years. The recorded recovery from drought-like conditions during these intervention periods is an indication that the interventions may have played a role in ushering more favourable weather conditions in all four districts, although this may also be a simple coincidence. Nevertheless, these results should be treated as important, and should be further explored. Natural resource management interventions appear to have successfully increased the adaptive capacity of the community during climate stress, especially in areas where the land and its holding capacity are more marginal. A large number of natural resource management activities are being implemented in the four concerned districts. The activities are typically aimed at managing and checking runoff from different catchments and reducing the sediment load in water bodies. The goal is to increase water resources and make the land more productive.

In Western Odisha, the capacity to adapt to changes in climate depends to a large extent on securing entitlements to natural resources, particularly water. Control over resources affects the strategies available, such as soil and water conservation, investment in resilient agriculture (pest or drought resistant seed, improved farming practices, etc.), and drawing on alternative sources of food and income when the main supplies fail. Water storage structures and soil and water conservation, developed in partnership with communities through participatory micro planning and the use of local labour, have provided more immediate ground water recharge, reducing intra-annual fluctuation in the water table and improving hydrological and soil moisture conditions. This has improved resilience to the increasingly variable monsoon rain, prolonged dry spells, and drought. Interventions have also had marked effects on groundwater tables, which have been raised by 2–4 metres.

Water harvesting technologies have also checked runoff and reduced sediment, increasing crop production and the productivity of water resources, which now include fish farming. Through increased water availability, the land use patterns have changed, permitting a second crop during the *rabi* season. The gross cropped area has expanded and cropping intensity has increased in the rain fed areas. Increased water availability affected agricultural production and productivity, mainly through greater crop diversification and improved crop yields. Yield increases have been regularly recorded. Livelihoods have become more robust through diversification into livestock, aquaculture, horticulture, silviculture plantation, and activities such as honeybee-keeping and mushroom cultivation. The growing practice of seed exchange and onion storage has also increased returns. Fewer households now suffer "lean season" food deficit days, a decrease from 25% before the project to 5% in 2009; thus food security has improved. This has happened as a result of increased coping capacity and increased income. The increased income resulted from an increase in agricultural production and a diversification of livelihood activities. This includes support and training in artisan craft work, and the establishment and management of small businesses by the very poor, particularly by women and the landless. These marginal groups have increased access to employment and consumption credit.

ICAI independent evaluation (post project) of WORLP

Further, WORLP had also undergone a post project evaluation by Independent Commission for Aid Impact (ICAI) of the UK government during 2012. ICAI reports that

> "the project was successful and contains much of what we consider to be best practice in delivering UK aid. WORLP was well designed and innovative. It had clear and relevant objectives. Incomes have been increased and livelihoods have become more secure. Communities were involved

throughout and are now better able to respond to climate variability (both droughts and heavy rains)."
(ICAI 2013, 20, section 3.1)

The project was rated as a success based on a 2010 impact assessment study that reports evidence from a random sample of 15% of the project's watersheds and interviews with 4200 people in 300 villages. It compared WORLP core and additional watersheds with "control" watersheds, where the Government of Odisha implemented land and water management under its other programmes and there were no WORLP activities. DFID's financing model was appropriate. 85% of financial aid (62% of the total funding which is £20.4 million) was transferred directly to beneficiaries as wages for work or grants and loans for community-based businesses. Only 6% of the financial aid was used to establish and fund OWDM systems (£2 million) and 11% (£3.6 million) to fund the local expertise that supported communities (the Project Implementing Agencies [PIAs]). These activities have continued after the project and are now funded by the Government of Odisha or through national programmes. This transition was facilitated by WORLP being aligned closely with government systems.

The project has achieved significant benefits for the communities in core watersheds it was designed to assist. Seventy per cent of these watersheds show improvements in agricultural productivity that beneficiaries directly attribute to WORLP. All project-funded water management structures are continuing to deliver benefits, two years after the project ended. The state government took full ownership of the project and the OWDM became the basis for a rapid expansion of community-led watershed development throughout the state of Odisha. The state government assessed WORLP's approach to be a success. As a result, in 2007, the Chief Minister of Odisha sanctioned an expansion of WORLP's approach into six other districts through a new programme, *Jeebika*. This was completely funded by the Government of Odisha and had a total value of £10 million. In addition, WORLP had contributed to the development of other projects in India as well as national guidelines for Watershed Management.

The project's estimated rate of return is 25.4% based on cost-benefit analysis. Since the methodology underestimated improvements in household incomes by as much as 75%, project impacts may have been higher than indicated in its reporting. The positive internal rate of return of the whole project and the positive impact on target beneficiaries in the core watersheds appears to indicate that, overall, it provided value for money. This judgement is reinforced by considerable subsequent investment in similar projects enabled by WORLP's demonstration effect. WORLP was one of the two DFID-financed rural livelihoods projects that significantly influenced the Government of India in developing its national guidelines. In particular, it influenced the design and operation of the £2 billion National Watershed Management Programme, implemented in 27 Indian states. WORLP's specific contributions have been how to deliver projects with the poorest to improve incomes and how to manage operations at state level. The project demonstrated a successful model of addressing poverty in Odisha. This led to the state government rapidly increasing its investment in supporting livelihoods through watershed development in 2007. Consequently, OWDM is now responsible for investments of over £200 million across all 26 districts in the state (activities funded by both national and state governments).

The link between the project's approach to reduce vulnerability and adaptation to climate change shows that there are opposing forces which interact and influence livelihoods. There are the positive effects of poverty reduction through project investment, and negative pressures exerted by climate change on an already fragile environment. A comprehensive impact assessment study of WORLP (WORLP 2011) reveals that approximately 29.2% of households who were earlier living below the poverty line are now earning more than the level of income desired to live above poverty line, which is significantly higher than the control population.

This reduction in poverty by diversifying livelihoods offers a good platform for climate change adaptation. WORLP has succeeded in reducing poverty, and this is a prerequisite in the development of greater climate change adaptability. WORLP as a project has strategically identified the poor and most vulnerable households through participatory well-being ranking. Further, these poor households have been organised into self-help groups. This has led to building institutions of the poor in watershed villages. Over 4254 Self Help Groups (SHGs) with around 65,000 members have been organised and strengthened during the project and continue to function well. The number and strength of SHGs has increased social cohesion, reduced people's vulnerability, and increased the opportunity for collective action in case of climate-related shocks. Groups are better able to manage common property resources, and provide quicker, better informed, and more appropriate responses to stress situations. In particular, the status and voice of women have been benefited through SHG activities.

In Western Odisha the capacity to adapt to changes in climate depends to a large extent on securing entitlements to natural resources, particularly water. Control over resources affects the strategies available to people for dealing with climatic change, such as soil and water conservation, investment in resilient agriculture (such as pest or drought resistant seed, improved farming practices), and the ability to draw on alternative sources of food and income when the main supply fails. The implementation of watershed works by Users' Groups and Watershed Development Committees has enhanced the entitlement processes in the community. The project has also ensured an overall decrease in the incidence of lean season food shortage days from among 25% of the poor to approximately 5%. The dual focus on natural resource management and people's livelihoods provides a strong response to increased vulnerability. Almost 86% of the marginal farmers in the project watersheds had reported improvement in disaster coping capacity. Enhanced agriculture production, diversification of livelihood activities, and access to consumption credit have been identified as the key factors behind improvement in disaster coping.

Conclusions

- **The project has proved to be sustainable**. In the core watersheds, water management structures are still functioning and are generally being well maintained by communities. As a result, the higher incomes achieved during the project are being sustained. More could have been done, however, to build the capacity of community organisations and link villages sustainably to market opportunities.
- **The project was clearly targeted to the needs of the poorest** in the four districts of Odisha where it was implemented. It improved the production of food and incomes. It reduced vulnerability to environmental and economic shocks. At the same time, it built communities' capacity to organise, resulting in the better management of land, assets, and production. Such improved organisation was particularly relevant for the most marginalised groups and the landless.
- **The project transferred resources directly to poor men and women**. Sixty-two per cent of the project's £32.75 million budget was transferred directly to beneficiaries. This was done as cash for work on watershed assets such as ponds. Funds were also transferred as grants or loans to build economic productive capacity. The other 38% of the funds strengthened expertise and organisations that proved to be largely sustainable. These funds significantly safeguarded the effectiveness and efficiency of the aid transferred to communities and reduced the risk of corruption.
- **WORLP involved beneficiaries in decision-making**. Those benefiting from UK funds were involved in decisions about how the money should be spent and were aware of

what they were meant to receive and why. Facilitation of this by locally based Indian NGO and government officials proved to be vital. This took time but was worth it in terms of impact and sustainability. Equally important was the full transparency of decision-making, record-keeping, and the management of project resources in villages and watersheds.

- **The project demonstrated effective partnerships**. This project demonstrated a close working relationship between the Governments of the UK, India, and the state of Odisha from the very beginning. DFID invested time and effort in building and maintaining this. The partnership was evident at senior and operational levels. DFID also actively sought to bring expertise into the project from NGOs in India and to work with them, particularly at local level.

- **Local and international resources were high quality**. This was fundamentally an Indian project, supported and facilitated by UK expertise and financing. The project is an example where DFID proved able to deliver high quality technical support through its staff and consultants over and above financial aid. As found in ICAI review of DFID's work in Bihar, DFID's skills and influence were highly valued by recipients and peer organisations.

- **The project's approach built on earlier learning about what works, combined with detailed local analysis in Odisha and in the focus districts**. It built on over a decade of lessons from DFID's support to rural livelihoods projects in India and experience from other programmes. It also built on global good practice. This combination meant that it was well designed.

- **The project was planned with an appropriately long timescale**. The project was designed to work over a 10-year period. It is not common practice in DFID to plan for projects over this length of time. The timing was, however, realistic given the nature of the programme. We note that DFID continued with the project in spite of some concerns about the pace of delivery, while community-level capacity was being built. This judgement proved to be correct.

- **There was continuity in leadership from both DFID and the Government of Odisha, as well as within the team of technical experts**. While there was some movement of individuals, this project demonstrates a higher level of continuity than is often seen. ICAI think this is an important factor in the project's success.

To summarise, WORLP in Odisha state of India sought to reduce poverty by improving communities' water resources, agriculture and incomes. It built infrastructure such as embankments, water storage ponds, and irrigation channels. It also provided loans and grants to the poor for community-based businesses. Incomes have been increased and livelihoods have become more secure. Communities were involved throughout and are now better able to respond to climate variability (both droughts and heavy rains). The Government of Odisha has taken full ownership of the project and the state and national governments have subsequently adopted approaches used by WORLP.

Notes

1. This zone is situated within 20° 9′ and 22° N latitudes and 82° 39′ and 85° 15′ E longitudes.
2. This zone is situated within 19° 3′ and 21° 55′ N latitudes and 82° 20′ and 83° 47′ E longitudes.

References

Behera, A. K., U. K. Mishra, B. C. Nayak, K. Das, B. Maharana, and P. C. Acharya. 2005. "Cropping System Strategy for Western Odisha Rural Livelihoods Project." APICOL, Bhubaneswar. WORLP working paper-52.

Census of India. 2011. http://www.censusindia.gov.in/2011census/PCA/PCA_Highlights/pca_highlights_ file/Odisha/Executive_Summary.pdf

DFID. 1999. *Sustainable Livelihoods Guidance Sheets*. London: DFID.

FAO. 2007. "Adaptation to Climate Change in Agriculture, Forestry and Fisheries: Perspective, Framework and Priorities".

ICAI. 2013. "DFID's Livelihoods Work in Western Odisha." (Report.18, February 2013).

Satyanarayana, M., Singha, A., Das, B., and Lopamudra. (2009). "Effect of Climate Change in Western Odisha". WORLP working paper-69.

WORLP. 1999. "Project Memorandum, Western Odisha Rural Livelihoods Project".

WORLP. 2008. "Impact Assessment Report, Western Odisha Rural Livelihoods Project." Sambodhi Research and Communications (Unpublished).

WORLP. 2011. "Impact Assessment Report, Western Odisha Rural Livelihoods Project." Sambodhi Research and Communications and Winrock International (Unpublished).

Environment and climate mainstreaming: challenges and successes

Emily Benson, Alex Forbes, Mika Korkeakoski, Razi Latif and Dechen Lham

This paper examines mainstreaming environment and climate change into development policy, planning, and budgeting. It looks at why we should integrate environment and climate and outlines challenges and successes. One result is that governments' progress pro-poor and equitable development. Governance gains are important too: co-benefits include more transparent decision making and better cross-government working. Ultimately, the impact of mainstreaming has increased awareness, changed perceptions, and improved the way inter-sectoral decisions are made, especially in climate adaptation. This supports countries to achieve their sustainable development ambitions – lessons which could be applied to a post-2015 development agenda.

Cet article examine l'intégration de l'environnement et du changement climatique dans les politiques, la planification et la budgétisation du développement. Il se penche sur les raisons pour lesquelles nous devrions intégrer l'environnement et le climat et présente les défis et les réussites. Un résultat est le fait que les gouvernements font des progrès vers le développement pro-pauvres et équitable. Les gains sur le plan de la gouvernance sont aussi importants : parmi les co-bénéfices figurent une prise de décisions plus transparente et un travail plus efficace entre tous les niveaux du gouvernement. En fin de compte, l'impact de l'intégration a favorisé la sensibilisation, modifié les perceptions et amélioré la manière dont les décisions intersectorielles sont prises, en particulier pour ce qui est de l'adaptation au changement climatique. Cela aide les pays à concrétiser leurs ambitions de développement durable – enseignements qui pourraient être appliqués à un ordre du jour pour le développement post-2015.

El presente ensayo examina la incorporación de los conceptos de medio ambiente y cambio climático en las políticas, la planeación y la financiación del desarrollo. Asimismo, analiza las razones por las cuales deben ser incorporados estos temas, señalando los retos enfrentados y los éxitos obtenidos. Resultan importantes los logros obtenidos por el gobierno, ya que los beneficios compartidos abarcan un proceso de toma de decisiones más transparente y una mejor coordinación entre dependencias gubernamentales. En última instancia, la incorporación de los temas mencionados ha elevado el grado de conciencia al respecto, ha modificado opiniones y ha mejorado la manera en que se toman decisiones a nivel intersectorial, particularmente en el ámbito de la adaptación al cambio climático. El proceso descrito podrá servir de apoyo para que los países logren sus planes orientados al desarrollo sostenible y, además, los aprendizajes obtenidos podrán ser integrados en una agenda de desarrollo pos-2015.

CLIMATE CHANGE ADAPTATION AND DEVELOPMENT

Introduction

The aim of this article is to demonstrate how the Poverty and Environment Initiative (PEI) has set out to increase awareness, change perceptions, and improve the way inter-sectoral decisions are made, especially in climate adaptation. The paper has been developed from the experiences of the PEI, a joint UNDP and UNEP programme implemented in 28 countries in Africa, Asia, Europe/ CIS, and Latin America. PEI is supported by the governments of Belgium, Denmark, Germany, Ireland, Norway, Spain, Sweden, the United Kingdom, the United States, and the European Commission, together with local and international think-tanks. Case studies for this paper have been drawn from *Stories of Change from the Joint UNDP-UNEP Poverty-Environment Initiative* (UNDP-UNEP Poverty-Environment Initiative 2013).

PEI is a catalytic programme that seeks to put in place enabling conditions (policies, instruments, capacities, and behaviours) that support the integration of environment sustainability into development processes, principally at the central, sector, and local government levels, but also within private sector and civil society institutions. The aim of PEI is to advocate to governments and development partners to provide longer-term support for a sustained increase in investments and capacity building in sustainability that is focused on lifting people out of poverty. This also contributes to the achievement of relevant national development priorities. PEI seeks to do this by changing perceptions and to demonstrate that investments in environmental sustainability can lead to reduced poverty and improved livelihoods.

What is environment and climate change mainstreaming?

At its most basic, environment and climate mainstreaming can be described as the integration of environmental and climate change objectives into non-environmental sectors and the "*greening*" of public policies (Hamdouch and Depret 2010; UNDP-UNEP Poverty-Environment Initiative 2011; Nunan, Campbell, and Foster 2012; Gazzola 2013).

Environment and climate mainstreaming has been defined as "*the informed inclusion of relevant environmental and climate change concerns into the decisions of institutions that drive national, local and sectoral development policy, plans, rules, investment and action*" (Dalal-Clayton and Bass 2009, 11). PEI recognises that climate change impacts:

> "cut across economic sectors, geographic and administrative boundaries and time scales. Therefore it is essential that (climate change) adaptation policies are formulated as part of broader policies for development." (UNDP-UNEP PEI 2009, 10)

The PEI defines mainstreaming as:

> "The iterative process of integrating poverty-environment linkages into policymaking, budgeting and implementation processes at national, sector and subnational levels. It is a multi-year, multi-stakeholder effort that entails working with government actors (head of state's office, environment, finance and planning bodies, sector and subnational bodies, political parties and parliament, national statistics office and judicial system), non-governmental actors (civil society, academia, business and industry, general public and communities, and the media) and development actors." (UNDP-UNEP PEI 2009, 6)

Many of the countries that are experiencing the greatest shocks due to climatic changes are low-income countries. In these countries, improved environmental management can reduce the impact of and improve recovery from extreme weather events (IPCC, WMO, and UNEP 2012). Climate change mainstreaming (Schaar 2008) follows in similar principle to environment mainstreaming whereby the challenge is to increase decision makers' awareness of climate change and variability, identify the aspects of national economies and population segments that are most sensitive to current risks and vulnerabilities, and build national capacity for improving the analysis of future risks and potential adaptation and mitigation strategies (World Bank 2009).

Mainstreaming displays, among others, the following characteristics (Nunan, Campbell, and Foster 2012; UNDP-UNEP Poverty-Environment Initiative 2013):

(1) it is an intentional process;
(2) many outputs can be targeted;
(3) it draws on and offers an interface between science-based analysis and policy decision-making;
(4) it requires a multi-disciplinary approach covering economic, social, environmental, and political disciplines;
(5) it takes place across multiple levels of government; and
(6) it guides implementation at central and local levels of government.

In addition, recent work being undertaken looks at mainstreaming and the theory of change, insofar as attempting to measure how the impact of mainstreaming programmes is going beyond simply policies and plans, but capturing that strengthened capacities and planning frameworks manifest behaviour change in institutions.

The development of mainstreaming, and "why mainstream?"

The ultimate goal of pro-poor environmental mainstreaming is to support sustainable and more equitable development. The Rio+20 meeting in 2012 signalled a fresh determination to deliver on the promise of over two decades of sustainable development. The summit's outcome document *The Future We Want*, underlines that reducing poverty is at the centre of a socially, environmentally, and economically sound world.

Sustainable development depends on successfully integrating the environment into economic planning and decision-making. Early efforts in the 1990s to mainstream the environment into national planning, for example through poverty reduction strategy papers (PRSPs), aimed to ensure that economic decisions, policies, and plans took environmental priorities into account and addressed the impact of human activities on environmental services and assets. Evidence suggested that these initial attempts to mainstream the environment into national planning had little success: a series of reviews by the World Bank showed that most of the PRSPs adopted by the world's poorest countries in the 1990s did not sufficiently address the environment's contribution to poverty reduction and economic growth (Bojö and Reddy 2003; Bojö et al. 2004).

A different approach for effective mainstreaming was needed, and evidence was emerging to support a new approach: the 2005 Millennium Ecosystem Assessment (MA) assessed the consequences of ecosystem change for human well-being and the scientific basis for action needed to enhance the conservation and sustainable use of those ecosystems and their contribution to human well-being and economies. The MA found that human actions are depleting the earth's natural capital, putting such strain on the environment that the ability of the planet's ecosystems to sustain future generations can no longer be taken for granted. At the same time, the assessment showed that with appropriate actions it is possible to reverse the degradation for many ecosystem services over the next 50 years, but the changes in policy and practice required are substantial (MA 2005). The Economics of Ecosystems and Biodiversity (TEEB) (2008 to date) builds on the MA outcomes by highlighting the economic impacts of ecosystem and biodiversity loss and the economic opportunities from recognising and responding to the economic values of ecosystem services and biological resources. UNEP's Green Economy (2011) and the Rio+20 report

(*The Future We Want*) strengthen the links between ecosystems, environment, economy, and human well-being by advocating for a green economy that contributes "*to eradicating poverty as well as sustained economic growth, enhancing social inclusion, improving human welfare and creating opportunities for employment and decent work for all, while maintaining the healthy functioning of the Earth's ecosystems*" (Rio+20 2012, Article 56, 9).

Today we see a growing body of evidence that suggests the failure to protect our environmental systems is undermining much of the progress that has been made in helping the world's poorest communities. Helen Clark, the UNDP Administrator, noted at a meeting in Costa Rica in March 2013 that "*the relative lack of success on MDG 7 relates in no small part to the failure to make clear links between ecosystem integrity, poverty eradication, and equity*" (Clark 2013, Final Para).

The ultimate goal of pro-poor environmental mainstreaming therefore is to improve people's lives *and* to improve the environment. This is achieved through making decisions differently. The goal of mainstreaming is to equip policymakers and planners with better evidence and tools which will improve both the way in which decisions are reached, and the efficacy of their decisions. Mainstreaming therefore focuses *on the way by which* people's lives and the environment are improved. The rationale is that by accelerating these actions, development goals are realised more quickly, people can adapt faster to a changing climate, and degradation can be halted or even reversed.

Mainstreaming challenges

Institutions and leadership

One of the immediate challenges for an environment or climate mainstreaming programme is finding the right national "institutional home". Quite often, as "environment" appears in the title, governments propose the ministry of environment as the lead agency. However PEI's experience indicates that pro-poor environmental mainstreaming across national policies, sector plans, and budgeting processes gains more traction and success if led and coordinated by overarching and powerful ministries of planning and/or finance. The challenge is to convince ministries of planning, local development, or finance to take the lead. Finding the right role then for ministries of environment (and other natural resource-related institutions such as water, land, and agriculture) in the mainstreaming process is also a key issue (see below).

Once a lead agency has been identified, defining pro-poor environmental mainstreaming in the context of current development issues, economic planning cycles, and political realities is often the next challenge. Ministers want solutions for tackling the issues of the day, rather than processes which bear fruit in years to come: it is a challenge to link a short-term need with a solution which provides benefit over years rather than months. Experience has shown that centring mainstreaming around themes or issues governments are currently grappling with will often find traction. Examples include waste management in Uruguay, natural resource extraction in Lao PDR, the high cost food imports in Malawi, and climate change in Bangladesh.[1]

After lead agencies and themes are agreed, links on the basis of evidence between poverty-environment-climate need to be made. The difficulty here is that while we know what the issues are, we may not have adequate evidence to reinforce the importance of issues and bring the linkages to the fore. A key issue at this stage is defining the problem and designing the analysis. This sometimes requires data that does not exist, and/or adapting existing data as well as innovative methodologies such as a "*climate public expenditure and institutional review*" (Bird et al. 2011, 2).

Experience has shown that finding the right person, a champion who has both gravitas and respect from government and the wider audience, to lead the study often results in greater traction and impact from the study's findings. In addition to this, designing how the study will be presented, who presents, the right audience, publicity around the study, etc., are all contributing factors to how well mainstreaming programmes will be taken up by governments.

Evidence from PEI has shown that often strong leadership drives mainstreaming and getting this leadership at the right hierarchical level to drive change is critical. Timing is also important in terms of releasing and communicating evidence and influencing planning and budgeting processes; for example, just as governments are about to embark on a new budget or five year economic plan.

Demonstrating and sustaining the results of mainstreaming is a significant challenge because changes that follow on from mainstreaming can occur slowly: legal and policy change takes time, as do increases in budgets or an agreement for a new institutional landscape. Sustaining results relies on national systems and ideally this should become an embedded government-wide process. A characteristic of many PEI country programmes is that personalities often drive and provide leadership to mainstreaming processes. Finding successors to replace mainstreaming champions is often difficult, as is maintaining the impetus and momentum of such processes.

Funding

Funding has been widely held to be a limiting factor, but PEI's experience has been that because mainstreaming is effectively a lobbying exercise backed by evidence and analysis, it is an inexpensive and much can be achieved with relatively modest amounts of money. PEI has found that once governments realise the value of mainstreaming, it becomes part and parcel of government's own work and hence can become a zero-external cost.

Implementation, implementation, implementation

The logical conclusion of PEI's work is that improved policies and plans are implemented and this yields development results more quickly and effectively. A challenge for PEI is to follow mainstreaming through to the "end of the pipe" which would look like improvements in the environment, livelihoods, and reduced poverty. For example, what are the impacts of changes in policies, institutions, and capacities for people on the ground? This hinges on government's capacity to deliver because ultimately taking mainstreaming forward is part and parcel of government's day-to-day work. Practically this could include poverty and environment indicators in the government's national poverty monitoring system; improving the budgeting process; and tracking the impacts of increases in budgets.

The ministries of local development (or their equivalent) have a mandate to coordinate, support, and advocate for decentralisation for sustainable, efficient, and effective service delivery. These ministries are increasingly important institutions to support cross-sectoral delivery by public institutions and in partnership with civil society and private sector institutions. In Bhutan, Lao PDR, Mozambique, Nepal, Rwanda, Tajikistan, Tanzania, and elsewhere, there has been increasing success in integrating pro-poor environmental sustainable objectives into sub-national development planning, implementation, and monitoring systems. The sub-national level can offer the comparable advantage of being more closely attuned to, and having better knowledge of the development needs of local communities and local stakeholders. This level of government and implementation is widely recognised as the most direct and effective area for development interventions and also as the most appropriate level for planning and implementing climate change adaptation measures.

Monitoring and evaluation

A results framework for PEI was developed in 2007 which details the intended outcomes, outputs, and corresponding indicators for monitoring and reporting on the progress of PEI. Key outcomes expected from PEI include enabling conditions, policies, instruments, capacity, and behaviours that support the integration of poverty-environment issues, principally at central and local government levels, but also within private sector and civil society institutions.

Consequently, PEI reflects an assessment of progress and achievements against enabling conditions and fostering institutional demand rather than aiming to assess PEI in relation to poverty reduction or environmental management impacts on the ground, which are both a product of many different actors and factors (recall that mainstreaming works on *the way by which people's lives and the environment are improved*). Quantifying the level of contribution (and corresponding attribution) that an enabling programme has had on measurable impacts such as a reduction in poverty and an improved environment is, at the moment, complicated to measure. One way PEI has tried to address this challenge is through establishing poverty-environment specific indicators in national monitoring systems; however, it remains a challenge to periodically collect information against the indicators by the national governments.

Lastly, a point to make in this section is that the mainstreaming described in this article is one that largely takes place between the UN and different layers of government. To be truly representational and more effective, it should involve poor and marginalised groups, and more active involvement from civil society organisations and the private sector.

Mainstreaming successes

Government

A key area which marks out PEI as different from other projects is that PEI specifically targets non-environment ministries. On paper the approach seems relatively straightforward but practice suggests it is far from simple. One of the key successes of PEI is that it stimulates government's own demand for environmental mainstreaming rather than offering to carry this out through an external project. By using economic and financial data presented in the language of planners and economists, this evidences how good environmental management can meet the wider development goals of government. These planners and policymakers then demand services from line ministries such as environment – which links back to a challenge outlined above – bringing ministries of environment into the mainstreaming process. PEI's experience has found that generating government's own demand for working in a more coordinated manner is a more effective and sustainable way to bring ministries such as environment into mainstreaming processes.

Governments working with PEI use the discipline of economics to better understand and communicate poverty-environment challenges. These include cost-benefit studies, expenditure reviews, and quantifying the value of natural and social capital. This marks an important shift: rather than relying on general arguments for sustainability or inclusive policy approaches, ministries have been able to quantify the costs and the benefits of different investment choices in a currency that the whole government understands.

Because mainstreaming is a cross-government activity, there is a need to coordinate between multiple ministries and departments. Climate change is one issue that has system-wide implications, ranging from the impacts on health and education, to infrastructure and disaster preparedness, and requires governments to make new connections across different parts of the system. Through a CPEIR, the government of Bangladesh realised that it spends US$1billion a year (6–7% of its annual budget) on climate change adaptation, where the popular assumption was that international donors were largely funding this. In a country where two-thirds of people depend directly on environmental resources for their livelihood, the issue of climate change is key for voters, who now know how much and how effectively their taxes are being spent by the government on responding to climate change. This has also helped to change the way the government in Bangladesh works: levels of cooperation between ministries have increased because the Ministry of Finance now recognises that the government's expenditure on climate change is one of national economic importance.[2]

In 2013, a Nepal study on Environmental causes of Displacement (NPC 2013) triggered the interest of the country's National Planning Commission (NPC) to better understand the links between migration and environmental stress, and, in particular, the issue of water shortages across many mountain districts. The study revealed how environmental stress can cause displacement of people who depend on ecosystem services for livelihoods in five selected districts prone to floods, drought, or landslides, and points out that many families are forced to migrate due to water shortages, floods, and landslides, without any hope of returning. Following the dissemination of the study findings and a site visit to one of the project areas, the NPC allocated 250 million rupees (approximately US$2.5 million) to a new programme to address the drought hit areas with the aim to reduce displacement due to environmental causes. This is an example of how a government has prioritised pro-poor development.[3]

Methodology

Mainstreaming an issue into government processes is not new – for example, gender mainstreaming has taken place in some countries. What's new in PEI is that it is a truly joint UN programme, making it easier to bring different strands of expertise together in a more coordinated fashion; PEI's architecture links national, regional, and global teams together; it focuses its efforts on non-sector ministries expressing poverty-environment opportunities (and costs) in an economic language which decision makers understand; PEI works to improve coordination on poverty, climate, and environment issues across governments; and in addition to changing policies, PEI also focuses on influencing budgets and financing.

The economic lens has helped governments convince their fellow decision makers and their electorate of the necessity for change. For example, the Government of Rwanda quantified the economic value of their own natural resources. On discovering that soil erosion, by the degradation of wetlands and forests, was forcing up the cost of electricity for communities by a staggering 167%, the government brought in a new programme for implementing sustainable farming techniques and supporting new livelihoods.

As a result, wetlands have now been restored and water levels are back to original levels. For Rose Mukankomeje, the Director General of the Rwanda Environment Management Authority (REMA), "*the results of the study were instrumental in the analysis of existing planning mechanisms and facilitated the identification of priorities within the environmental sector*". For her, the experience of working across different departments and drawing on different areas of expertise was also a "*clear demonstration that true partnership breeds success*" (UNDP-UNEP 2013, 54).[4]

Similarly in Malawi, the costs and benefits of sustainable and unsustainable natural resource management were quantified and compared in four areas: forestry, fisheries, wildlife, and soils. Findings showed that unsustainable natural resource use and management costs the country the equivalent of 5.3% of GDP annually. Soil erosion alone reduces agricultural productivity by 6%. The work of Benin et al. (2008), who use the IFPRI computable equilibrium model, indicated that recovering 6% growth in agricultural yields during 2005–15 would increase overall GDP growth from 3.2 to 4.8% per year leading to the proportion in poverty falling to 34.5% by 2015, i.e., an additional 1.88 million people being lifted above the poverty line by 2015. The economic analysis not only demonstrated the macro-economic contribution of natural resources to GDP but showed the links between investing in ecosystems and poverty alleviation. This work has contributed to a marked shift in the way that government institutions understand the issues.[5]

The UN

It is not uncommon to hear that UN agencies, like their government counterparts also suffer from fragmentation and a lack of coordination. The UNDP and UNEP partnership for PEI represents a good example of inter-agency cooperation within the UN system through joint decision-making, joint funding, and joint management.

Why? Because the PEI partnership is a more joined-up UN; other UN agencies look to the PEI as a good working model of a joint programme; and PEI represents a fresh approach where through effective mainstreaming, development outcomes are achieved more quickly.

The collaboration between UNDP and UNEP has proved especially successful at the country and regional levels, where programmes are designed and implemented through joint teams. At the country level, while PEI primarily works through the UNDP Country Office, PEI is now engaging with the wider UN Country Team to (1) demonstrate a joint programme in action; and (2) to canvas further support for poverty-environment-climate mainstreaming.

Evaluations of UNDP, UNEP, and PEI have highlighted this partnership is a successful area of work and a model of how UN agencies can work better together. PEI is scaling up the UN partnership; for example, in the Asia Pacific region, the United Nations Capital Development Fund (UNCDF) has joined PEI.

What next?

PEI is helping to improve the way governments do business by stimulating strengthened vertical policy, planning, and budgeting processes between national and sub-national institutions, and horizontal cross-sectoral initiatives that aim to contribute to pro-poor environmental sustainability. It should be noted that this is a long journey where challenges of supporting implementation and better measurements of impact still loom. A key issue is ensuring that changes to national plans and impacts from new sectoral priorities start to reach people on the ground. This will require continued coordination across the multitude of different activities going on at the national, sub-national, and local levels.

An opportunity exists for how lessons from mainstreaming can input into the post-2015 development agenda. Mainstreaming experience has demonstrated that many MDGs are interlinked, and that achieving the goal on poverty reduction will be not possible if other related goals, including environmental sustainability, are not met. Newly emerging ideas of measuring human and environmental dimensions of development and going beyond measurements of GDP are gaining traction internationally. Experience from Bhutan has shown that quality of life can be expressed beyond the measure of GDP per capita and that the integration of social and environmental factors are equally important in ensuring wider development goals are reached.

More exciting, and going beyond government-led mainstreaming, is that green jobs, social enterprises, and wealth accounting are among many other initiatives which are increasingly being seen as ways to address the integration of true social and environmental costs for a better quality of sustainable development.

Overall, a key challenge for PEI is establishing a clear link between PEI interventions in areas such as natural resource management and their impact on poverty reduction. Some work, like the PEI Malawi economic analysis, has shown positive results, but more work needs to be done – establishing this direct link remains problematic for all policy and capacity building programmes.

Notes

1. For PEI country profiles, brochures, and reports see UNDP-UNEP Poverty and Environment Initiative. Accessed November 11, 2013. http://www.unpei.org/what-we-do/countries
2. For more information on Bangladesh see UNDP-UNEP Poverty and Environment Initiative. PEI Bangladesh Factsheet. Accessed November 11, 2013. http://www.unpei.org/what-we-do/pei-countries/bangladesh. This page contains information and reports from the PEI and related programmes in Bangladesh.
3. For more information on Nepal see UNDP-UNEP Poverty and Environment Initiative. PEI Nepal Factsheet. Accessed November 11, 2013. http://www.unpei.org/what-we-do/pei-countries/nepal. The video 'Local governments go green in Nepal' is a good overview of PEI. Also see the 'Influencing Policy Processes' series of publications under 'Key Documents'.
4. For more information on Rwanda see UNDP-UNEP Poverty and Environment Initiative. PEI Rwanda Factsheet. Accessed November 11, 2013. http://www.unpei.org/what-we-do/pei-countries/rwanda Under key documents see 'Economic Analysis' and 'Integrated Ecosystem Assessment'.
5. For more information, see the report 'Economic Analysis of Sustainable Natural Resource Use in Malawi'. Accessed November 11, 2013. http://www.unpei.org/what-we-do/pei-countries/malawi (under 'Key Documents' and 'Economic Analysis').

References

Benin, S., F. Byekwaso, E. Kato, M. Kyotalimye, G. Lubadde, E. Nkonya, G. Okecho, and J. Randriamamonjy. 2008. *Impact of Uganda's National Agricultural Advisory Services Program.* International Food Policy Research Institute.

Bird, N., T. Beloe, M. Hedger, J. Lee, M. O'Donnell, and P. Steele. 2011. *Climate Public Expenditure and Institutional Review: A Methodology to Review Climate Policy, Institutions and Expenditure.* Manila: Asia-Pacific Capacity Development for Development Effectiveness Facility with UNDP and ODI.

Bojö, J., and R. C. Reddy. 2003. *Status and Evolution of Environmental Priorities in the Poverty Reduction Strategies: An Assessment of Fifty Poverty Reduction Strategy Papers.* World Bank Environment Department Paper no.93. Washington, DC: World Bank.

Bojö, J., K. Green, S. Kishore, S. Pilapitiya, and R. C. Reddy. 2004. *Environmental in Poverty Reduction Strategies and Poverty Reduction Support Credits.* World Bank Environment Department Paper no.102. Washington, DC: World Bank.

Clark, H. 2013. "Opening remarks at the Post-2015 Environmental Sustainability Consultation on 'Linking Poverty Eradication, Equity, and Environmental Sustainability in the Post-2015 Global Development Agenda'". Accessed March 25, 2014. http://www.undp.org/content/undp/en/home/presscenter/speeches/2013/03/18/helen-clark-opening-remarks-at-the-post-2015-environmental-sustainability-consultation/

Dalal-Clayton, B., and S. Bass. 2009. *The Challenges of Environmental Mainstreaming: Experience of Integrating Environment into Development Institutions and Decisions.* Environmental Governance No. 3. London: International Institute for Environment and Development.

Gazzola, P. 2013. "Reflecting on Mainstreaming through Environmental Appraisal in Times of Financial Crisis — From 'Greening' to 'Pricing'?" *Environmental Impact Assessment Review* 41: 21–28.

Hamdouch, A., and M.-H. Depret. 2010. "Policy Integration Strategy and the Development of the 'Green Economy': Foundations and Implementation Patterns." *Journal of Environmental Planning and Management* 53 (4): 473–490.

IPCC, WMO, and UNEP. 2012. *Managing the Risks of Extreme Events and Disasters to Advance Climate Change Adaptation – Special Report of the Intergovernmental Panel on Climate Change.* Geneva: IPCC, WMO, and UNEP.

Millennium Ecosystem Assessment (MA). 2005. *Ecosystems and Human Well-being: Synthesis*. Washington, DC: Island Press.

National Planning Commission. 2013. Environmental Causes of Displacement, Published by Government of Nepal, National Planning Commission with support from UNDP/UNEP in Kathmandu, Nepal.

Nunan, F., A. Campbell, and E. Foster. 2012. "Environmental Mainstreaming: The Organisational Challenges of Policy Integration." *Public Admin. Dev.* 32: 262–277.

Rio+20. 2012. "The Future We Want." Rio+20/UN website. https://www.un.org/en/sustainablefuture/

Schaar, J. 2008. *Overview of Adaptation Mainstreaming Initiatives*. Stockholm: Commission on Climate Change and Development.

UNDP-UNEP Poverty-Environment Initiative. 2009. *Mainstreaming Poverty-Environment Linkages into Development Planning: A Handbook for Practitioners*. Nairobi: UNDP-UNEP Poverty-Environment Facility.

UNDP-UNEP Poverty-Environment Initiative. 2011. *Mainstreaming Climate Change Adaptation into Development Planning: A Guide for Practitioners*. Nairobi: UNDP-UNEP Poverty-Environment Facility.

UNDP-UNEP PEI 2013. *Stories of Change from the UNDP UNEP Poverty and Environment Initiative*. UNDP-UNEP Poverty-Environment Facility.

World Bank. 2009. *Policy and Institutional Reforms to Support Climate Change Adaptation and Mitigation in Development Programs*. Washington, DC: World Bank.

Index